The Fate of Nutrients and Pesticides in the Urban Environment

ACS SYMPOSIUM SERIES **997**

The Fate of Nutrients and Pesticides in the Urban Environment

Mary T. Nett, Editor
Water Quality Consulting

Mark J. Carroll, Editor
University of Maryland

Brian P. Horgan, Editor
University of Minnesota

A. Martin Petrovic, Editor
Cornell University

**Sponsored by the
ACS Division of Agrochemicals**

American Chemical Society, Washington, DC

ISBN 978-0-8412-7422-8

PRINTED IN THE UNITED STATES OF AMERICA

Foreword

The ACS Symposium Series was first published in 1974 to provide a mechanism for publishing symposia quickly in book form. The purpose of the series is to publish timely, comprehensive books developed from ACS sponsored symposia based on current scientific research. Occasionally, books are developed from symposia sponsored by other organizations when the topic is of keen interest to the chemistry audience.

Before agreeing to publish a book, the proposed table of contents is reviewed for appropriate and comprehensive coverage and for interest to the audience. Some papers may be excluded to better focus the book; others may be added to provide comprehensiveness. When appropriate, overview or introductory chapters are added. Drafts of chapters are peer-reviewed prior to final acceptance or rejection, and manuscripts are prepared in camera-ready format.

As a rule, only original research papers and original review papers are included in the volumes. Verbatim reproductions of previously published papers are not accepted.

ACS Books Department

Contents

Indexes

Preface

Although often taken for granted, the presence of turfgrass has come to define our nation's urban and suburban landscapes. The functional role of healthy turfgrass in the modern environment cannot be overstated, given its significant and positive impact on society's and the environment's well-being. The products available to homeowners and professionals to maintain healthy turf have been used for more than 50 years, with new technologies entering the marketplace each season. Recently, pesticide and fertilizer products used on turfgrass have become the focus of public policies related to water quality. These policy decisions have frequently conflicted with recommendations developed by turf scientists to aid homeowners and turfgrass managers on the timing and amounts of water, fertilizers and pesticides needed to maintain a healthy turfgrass sward.

Given the importance of having the most current research available to inform public debate and support effective public policy, a symposium, *The Fate of Nutrients and Pesticides in the Urban Environment*, was held October 12–13, 2005 in Arlington, Virginia; the meeting assembled a group of the nation's leading turfgrass and environmental scientists to discuss the fate and transport of fertilizers and pesticides in the urban landscape. The goal of the workshop was to identify the realities of chemical transport in turf systems and to assess how appropriate maintenance practices performed on urban lawns may or may not impact water quality. The American Chemical Society (ACS) Division of Agrochemicals and the non-profit national trade association RISE (Responsible Industry for a Sound Environ-ment)© hosted the symposium, which brought together university andfederal government scientists, regulators, environmental advocates, and turfgrass managers.

This book represents a compilation of research on chemical movement in turf, or turf-predominate systems, having scales of measure ranging from small-sized plots to a single 332 hectare watershed. The research examines how turf management practices, basic physico-chemical turfgrass properties and regional weather patterns can affect the transport of nutrients and pesticides. A comparison with alternative landscapes is a frequent theme in several of the papers included in this book. Papers focusing on nutrient and pesticide leaching are accompanied by an equal balance of papers addressing nutrient and pesticide runoff from golf turf and home lawns. In nearly all cases, the rates and timing of nutrient and pesticide applications used within the studies presented within this book are consistent with recommendations typically provided by turf scientists to turfgrass professionals and on product labels created for homeowners. Thus, many of the conclusions drawn from the research presented in this book represent a likely characterization of the risks—or absence of risk—to water impairment that may be expected from turf areas receiving nutrients and pesticide inputs.

Acknowledgments

We gratefully acknowledge the book's contributors for their continued dedication to the science of turfgrass as well as for their excellent presentations at the symposium and commitment to producing the book. We extend special thanks to Mary Nett, who ably kept production on track, and to Novak Birch, Baltimore, Maryland, for creating the book's cover art.

And finally, thanks to the trade association RISE® and The ACS Division of Agrochemicals for bringing together the best turfgrass research available through their co-sponsorship of the symposium and book.

Jim Skillen

RISE (Responsible Industry for a Sound Environment)©
1156 15th Street NW
Suite 400
Washington, DC 20005
Email: jskillen@pestfacts.org

Chapter 1

Management Practices Affecting Nitrogen and Soluble Phosphorus Losses from an Upper Midwest Lawn

Wayne R. Kussow

Department of Soil Science, University of Wisconsin at Madison, Madison, WI 53706

Public and regulatory agency concerns about the impacts of lawns on surface and groundwater quality were addressed in this 6-year study. The research site was a Kentucky bluegrass lawn established on a 6% slope at the University of Wisconsin-Madison turfgrass research facility. The lawn stand density maintained was sufficient to prevent sediment loss. Under these circumstances, depths of runoff ranged from 30–74 mm, or 5–9% of annual precipitation. Runoff inorganic N and soluble P losses ranged between 0.08–1.57 and 0.031–1.29 kg ha^{-1}, respectively. Annual leachate nitrate-N concentrations were typically between 2 and 4 mg L^{-1} and the quantity of N leached averaged 3.9 kg ha^{-1}. These N and P losses are intermediate between those from naturally vegetated and agricultural areas in the upper Midwest. Subsoil compaction and topsoil manipulation prior to lawn establishment did not affect runoff depths or N and P losses. This was attributed to the fact that over the duration of the study, 87% or more of the runoff and N and P losses occurred during the period of December through March. The single factor most strongly relating to runoff loss of N and P was depth of runoff. Failure to fertilize the lawn was second in importance. Impacts of other cultural practices such as chisel plowing of the compacted subsoil and type of fertilizer applied were of lesser importance.

In the early 1990s, concerns began to surface regarding the impacts of managed turfgrass on surface and groundwater quality in the upper Midwest. There was little research from the region to refute or validate these concerns. Indications from research conducted in Maryland (*1*) and Rhode Island (*2*) were that lawn runoff depths might range from 1 to 14 mm with sediment loads of 2–25 kg ha^{-1} and N and P loads of 0.06 to 0.0084 kg ha^{-1}. Under proper management, leaching losses of N might approach 2 kg ha^{-1} and leachate nitrate–N concentrations might range from 0.35–4.02 mg L^{-1}. These values are in stark contrast to studies conducted earlier on field crop rotations in Minnesota (*3, 4*). Annual runoff averaged 86.3 mm over 10 years. Sediment loss averaged 3.960 kg ha^{-1}, annual runoff water N and P loads ranged from 1.07–1.15 kg ha^{-1}, tile drainage water averaged 12.7 mg L^{-1} of nitrate-N, and average leaching loss of N was 42 kg ha^{-1} yr^{-1}. In this era, watershed studies in Maine (*5*) recorded annual runoff water P loads of 0.14 kg ha^{-1} from a low density residential area and 0.019 kg P ha^{-1} from an adjacent forested area. A P load of 0.09 kg ha^{-1}yr^{-1} was reported for a Minnesota forest (*6*) and runoff losses from native prairies were reportedly in the range of 0.8 kg N ha^{-1}yr^{-1} and 0.1 kg P ha^{-1}yr^{-1} (*7*).

While these numbers provided evidence that the potential contributions of lawns to the deterioration of surface and groundwater quality in the upper Midwest are considerably less than for agricultural land and no greater than for forests and prairies, the contributions of lawns to urban runoff P loads remained an unknown. Indications were that, in this region, snowmelt is the major contributor of runoff and annual N and P loads (*3, 7*). This introduces desiccated vegetation as a significant source of P in runoff water (*8, 9, 10*).

The purpose of the present study was to assess the influences of management practices on N and P losses from an upper Midwest Kentucky bluegrass lawn. Design of the study took into account the obvious need to measure runoff throughout the year and to test allegations that soil compaction, fertilization practices, and type of fertilizer applied significantly impact N and P losses from lawns.

Materials and Methods

The study site was the lower one-half of a 62 m long, southeast-facing slope at the University of Wisconsin O. J. Noer Turfgrass Research and Education Facility. An eroded area had to be renovated prior to turfgrass establishment and installation of an automated irrigation system. Earth-moving equipment was used to strip and stockpile 15 cm of topsoil from the non-eroded area. A second cut of soil of comparable depth was likewise removed and stock piled. A bulldozer was then used to move soil over the eroded area. The second cut of soil was spread over the area, plots temporarily marked, and select plots compacted with a vibrating roller of the type used for road construction. One-third of the compacted plots were cultivated via a single pass with a chisel plow whose shanks were spaced 20 cm apart and were adjusted to till the soil to a 20 cm depth. One-half the stockpiled topsoil was spread over the plot area and select plots rototilled to a 15 cm depth. The remainder of the topsoil was replaced and the area graded to a target slope of 6%. A seedbed was prepared, the area seeded to a blend of four Kentucky bluegrass cultivars, and hydro-mulched.

Two months after seeding, an irrigation system was installed on plots measuring 2.4 x 9.2 m, bounded with plastic lawn edging that protruded 2.5 cm above the soil surface. Each plot was outfitted with runoff collectors (11) fabricated from 18-gauge steel. The exit chutes were outfitted with three-slot sample splitters calibrated to determine the fraction of total runoff actually exiting the middle slot. A chute from the middle slot conveyed the runoff water to a 10 L plastic pail inside a 105 L plastic trash can set into the ground. During the first winter after installation, freezing water forced the trash cans partially out of the ground. The problem was resolved by installing drain pipes under each can, imbedding the cans in pea gravel, and staking each can with 60 cm lengths of angle iron.

A modified version of a low tension capillary wick lysimeter (12) was installed in the center of each plot. Modifications included replacement of the glass plate with a 2 cm diameter plastic funnel, fanning out of the glass wick (Pepperell Braiding Co., Inc., Pepperell, MA) over the surface of a plexiglass insert rather than using a glass plate covered with glass cloth and using two rather than six strands of the 1.25 cm diameter wicking. The wicking passed through rigid plastic tubing heated to create an L-shape and ending in a 4 L sample collection bottle. The total length of the fiberglass wicking was 86 cm, providing a tension of –86 kPa.

Installation of the lysimeters began with removal of sod from a 60x60 cm area followed by soil excavation to a 120 cm depth and stockpiling the different layers of soil. A 30x30x30 cm cut was made into one sidewall, the top of the cavity being positioned 45 cm below the soil surface. After wetting the capillary wicks, the lysimeter funnel was forced up against the top of the cavity and moist soil packed around it. The layers of soil were replaced in their original sequence and the sod tamped back into place. Each lysimeter collection bottle was outfitted with plastic air entry and sampling tubes that terminated in plastic screw-cap containers whose tops were positioned 2 cm above the soil surface to provide easy access, but not interfere with mowing. The collection bottles were emptied employing a hand vacuum pump (Model 2005 GZ, Soil Moisture Corp., Santa Barbara, CA). The leachate collection bottles were emptied twice over a 2 month period before leachate volumes were recorded and subsamples collected for analysis. Leachate was thereafter collected monthly during the period extending from termination of snowmelt to the time of permanent snow cover. At no time did the volumes of leachate collected exceed the total volume of the collection bottles.

Volumes from all runoff events were recorded and subsamples collected for analysis. The subsamples were stored at $-4°C$ until analyzed for total inorganic N (13) and for soluble orthophosphate P (14). Detection of only soluble orthophosphate P was justified on the basis that at no time did the runoff water contain measurable quantities of sediment and that analysis of some of the first samples collected revealed that 95% or more of the total P was soluble orthophosphate P. Some samples were noticeably discolored with organic material, raising the possibility that they contained significant amounts of organic P that the employed procedure does not detect. Leachate orthophosphate P was periodically measured and found to never deviate significantly from a mean value of 0.03 mg L^{-1}.

In project years 1 and 2 (project year was May through April), the study focused on the pre-lawn establishment effects of soil manipulation on runoff. The treatments were non-compacted subsoil, compacted subsoil, and chisel plowed compacted soil, each with and without topsoil tilled into the subsoil. Absent any significant influences of soil manipulation on runoff depths or N and P losses (Table IV), the scope of the study was expanded by devoting the next two project years to investigation of the effects of fertilization practices and clipping management on runoff and N and P losses. The treatments were no fertilizer or application of synthetic (Scotts Turfbuilder® 29-3-4) or organic (Milorganite® 6-2-0) fertilizer, each with clippings removed or left on the lawn via mulch mowing. The treatment variable in the final 2 years of the study was the type of N carrier. Included were urea, a biosolid, polymer + sulfur coated urea (PSCU), polymer coated urea (PCU), methylene urea and isobutyldiene diurea (IBDU). Sources of the N carriers were agronomic grade urea with 45% N, Milorganite (6-0.88-0), Scott's Poly-S® 25-1.3-8.3, Turfgo® 25-2.2-8.3, Scotts 22-0-13.2 and ParEX® 31-0-0®, respectively.

When clippings were not removed, fertilization consisted of four 49 kg ha^{-1} N applications per season. Times of application were early May, July, September, and in late October after the plots were mowed for the last time that season. When clippings were not removed, the September application was eliminated, reducing to three the number of 49 kg N ha^{-1} applications per year. In project years 1 and 2, the fertilizer applied had a grade of 25-1.3-8.3 (N-P-K) and the N carrier was polymer + sulfur-coated urea. The plots were irrigated when a color change in the turfgrass signified the onset of moisture stress. The run time was 30 minutes, during which the plots on average received 9.6 mm water. Depending on wind direction and speed, water applied to the individual plots ranged from 5.8–12.4 mm. At no time did irrigation induce runoff. The plot mowing height was 64 mm, at a frequency dictated by grass growth rate. A single application of a three-component broadleaf herbicide was made each year in September.

The original soil on the site was a Troxel silt loam (fine silty, mixed, superactive, mesic Pachic Argiudoll). The top 15 cm averaged 20.3% sand,

55.7% silt, and 24.0% clay. At start-up time, the soil pH was 6.5, the organic matter content 2.1%, and the Bray P1 extractable P and K levels were 60 and 176 ppm, respectively. According to soil test interpretations in Wisconsin, both nutrient levels are three times those deemed adequate for established lawns. Other soil characterizations included bulk density and water infiltration rates. Soil bulk density was determined at 2.5 cm depth intervals from 25 cm long cores removed with a soil bulk density sampler (15). Two methods were used to determine water infiltration rates. One employed a double ring infiltrometer manufactured specifically for use in turf (Turf-Tec International, Coral Springs, FL). The second approach was that of applying simulated rain for 1 hour at the rate of 6.25 cm hr^{-1} and calculating infiltration rates using runoff volumes for the last 5 minutes.

The treatments employed in project years 1–4 could have been deployed in a split-plot (nested) experimental design. Instead, a randomized complete design was employed because of the concern that the split-plot design could become restricting in terms of treatments that might be selected for future investigations.

All data were statistically analyzed by way of a one-way ANOVA performed with COHORT software (COHORT Software, 798 Lighthouse Ave., PMB 320, Monterey, CA 93940). For comparison of treatment means, LSD values were computed at 5 and 10% probabilities. To check for subsoil-topsoil manipulation interactions during years 1 and 2 and fertilizer-clipping management interactions in years 3 and 4, the data were analyzed assuming a split-plot design. Indications were that no interactions arose in either set of treatments.

Results and Discussion

Experimental Site Characteristics

The 30 year norm for annual precipitation in Madison, WI is 764 mm. Project year (May through April) precipitation ranged from 641–922 mm (Table I), with variances of –16% to +21% from the 30 year norm, respectively. Variations of this magnitude emphasize the importance of multi-year measurement of runoff and nutrient losses.

Soil disturbance effects on compaction were less than anticipated. According to bulk density measurements (Table II), roller compaction of the subsoil increased bulk densities an average of only 5%, from 1.40–1.56 Mg m^{-3} at the 15 to 20 cm soil depth. An even smaller increase of 2.5% occurred at the 20 to 25 cm depth. Rototilling topsoil into the subsoil resulted in a 3% increase in the bulk density of the top 5 cm of soil. The combined effects of stripping and replacing the topsoil resulted in a mean bulk density of 1.48 Mg m^{-3}. This is an

Table I. Project Year Precipitation

Month	Project Year Precipitation						Monthly Mean
	1	2	3	4	5	6	
				mm			
May	16	80	38	113	80	52	63
June	26	158	174	40	203	151	125
July	171	53	93	140	109	154	120
Aug	99	91	38	79	21	50	63
Sept	143	89	108	44	28	36	75
Oct	46	18	24	111	85	27	54
Nov	170	24	40	45	29	28	56
Dec	61	3	16	23	23	11	23
Jan	16	6	173	34	21	28	46
Feb	4	43	2	10	36	50	24
Mar	58	15	57	14	36	118	50
Apr	112	61	128	52	64	117	87
Totals	922	641	891	705	735	822	786

increase of 11% over the bulk densities of 1.32 Mg m^{-3} commonly found for the plow layers of Wisconsin silt loam soils (16).

Water infiltration rates measured with the double-ring infiltrometer ranged from 4.26–4.93 cm hr^{-1} and there were no statistically significant differences among the soil treatments (Table II). For the non-compacted and compacted soil treatments, infiltration rates determined via rainfall simulation were nearly 2 cm hr^{-1} less than those measured with the double-ring infiltrometer, a difference to be expected given the fundamental differences in methodologies. Regardless, both sets of values fall within the 2.5–8.1 cm hr^{-1} lawn infiltration rates reported in other studies (17, 18).

Only the rainfall simulation approach detected an influence of chisel plowing of compacted subsoil on infiltration rates (Table II). A plausible reason for this is the differences in soil surface area encompassed by the double-ring infiltrometer and the metal frame used in the rainfall simulation method. Only by chance would the double-ring infiltrometer, with an outer ring diameter of 11 cm be positioned over one of the 20 cm spaced channels created by the chisel plow. In comparison, the 91x91 cm frame used in rainfall simulation would have traversed three to four chisel plow channels. Therefore the rainfall simulation derived infiltration rates are thought to have been more sensitive to the effects of chisel plowing. This leads to the implication that chisel plowing of compacted subsoil initially has the potential for increasing lawn infiltration rates by 42 to 130%, depending on whether the topsoil is layered or rototilled into the compacted subsoil.

A distinguishing characteristic of the upper Midwest as contrasted with warmer climates is the time of major runoff events and runoff losses of N and P. As shown in Table III, runoff and N and P losses occurred predominantly during

Table II. Soil Bulk Density and Infiltration Rates as Influenced by Soil Disturbance and Remedial Action

Soil Treatment[a]		Soil Bulk Density by Depth (cm)					Infiltration Rate	
							Double	Sim.
Subsoil	Topsoil	0-5	5-10	10-15	15-20	20-25	Ring	Rain[b]
		-------------------- $Mg\ m^{-3}$ ------------------					------ $cm\ hr^{-1}$ -----	
Not	Layered	1.46	1.43	1.50	1.47	1.58	4.75	1.93
	Mixed	1.50	1.42	1.48	1.48	1.58	4.75	1.96
Comp	Layered	1.44	1.38	1.50	1.57	1.62	4.57	1.40
	Mixed	1.49	1.43	1.46	1.56	1.60	4.27	1.78
Comp+	Layered	1.47	1.40	1.51	1.54	1.64	4.70	2.36
chisel	Mixed	1.51	1.44	1.44	1.55	1.60	4.83	4.31

[a] Not = not compacted; Comp = Compacted; Comp+chisel = Compacted plus chisel plow.

[b] Simulated rain at 6.8 cm hr^{-1} for 1 hour.

December through March. The percentages of annual runoff during this period ranged from 68.5% in project year 2 to 99.0% in project year 6, resulting in a 5 year mean of 87.5%. These percentages are comparable to the 62% reported for lawns in New York (19) and the 61 to 94% range observed in a 10 year study in Minnesota (3).

The 68.5% of total runoff measured for project year 2 (Table III) came in a year when total precipitation was 123 mm below the 30 year norm (Table I). The 99.0% of winter runoff in project year 6 was associated with a unique situation. Twelve separate runoff events resulted during the winter months from repeated temporary thaws and two rainstorms in March, leaving only a light snow covering for spring melt.

Percentages of annual runoff N and P loads collected from December through snowmelt paralleled those for runoff (Table III). Besides suggesting a strong relationship between runoff volumes and N and P losses, this observation raised questions regarding the sources of runoff N and P during the winter months. These runoff events occurred two or more months after the most recent fertilizer application and, with no sediment in the runoff water, frozen soil contributions of soluble P were presumably small. Based on other research (8, 9, 10) and the influence of turfgrass freezing and drying (Table IV), the evidence is that tissue P can be the primary source of winter runoff P from lawns.

Soil Treatment Effects

Due to heaving of the runoff collectors by ice during the winter of project year 1, that data set is incomplete. Consequently, the full year data being presented are restricted to project year 2. In this, the driest of the six project

Table III. Percentages of Annual Runoff and Annual Runoff N and P Collected December through March Snowmelt

Winter of:	Runoff	Runoff N	Runoff P
	---------------------- % of Annual Total ----------------------		
Project year 2	68.5	65.6	74.6
Project year 3	78.3	81.1	86.9
Project year 4	96.2	91.9	92.5
Project year 5	95.7	96.4	96.0
Project year 6	99.0	97.8	98.5
Mean	87.5	86.6	89.7

Table IV. Potential Contributions of Kentucky Bluegrass Tissue P to the Soluble P in the Equivalent of 25 mm Runoff Water

Fertilizer N Rate kg ha^{-1}yr^{-1}	Grass Biomass P^a kg ha^{-1}	Clipping Pre-treatment	Clipping P Percent	Leached^b Amount kg ha^{-1}
0	3.58	None (fresh)	3.7	0.013
		Air-dried	13.4	0.480
		Frozen/air-dried	15.8	0.566
196	8.05	None (fresh)	6.3	0.507
		Air-dried	22.9	1.832
		Frozen/air-dried	19.8	1.594

[a] Oven-dry weight basis.
[b] Five-minute shaking time.

years, annual runoff ranged from 30.1–38.9 mm (Table V), or 5 to 6% of total precipitation. Winter runoff contributed 68.5% of the annual total (Table III).

Annual runoff N loads never exceeded 0.177 kg ha^{-1} (Table V), an amount intermediate to those observed in Maryland (1) for 2 years when annual precipitation ranged from 828–1040 mm. Annual soluble P loads in the present study of 0.236–0.284 kg ha^{-1} were considerably greater than those observed in the Maryland study. Their average ratio of runoff N to soluble P was more than 17:1, very different from the 0.8:1 found in the present study, possibly indicating yet another distinguishing feature of runoff loss of N and P when winter runoff dominates. The runoff N to P ratio of 0.8:1 also differs markedly from the 19:1 N to P ratio in the fertilizer applied.

Soil disturbance treatment effects on amounts of runoff and N and P runoff losses were not significant at the 90 or 95% probability levels (Table V). While high variability in the data (coefficients of variation ranging from 30–81%) may be cited as the reason for lack of significant soil disturbance effects, time of runoff and N and P losses should not be ignored. Over 68% of the runoff

**Table V. Project Year 2 Runoff Water and N and P Loss as a Result
of Soil Disturbance Prior to Lawn Establishment**

Soil Disturbance[a]		Runoff Loss				Leachate	
Subsoil	Topsoil	Depth mm	N ---- kg ha^{-1}yr^{-1} ----	P	Depth mm	NO_3-N mg L^{-1}	N kg ha^{-1}yr^{-1}
Not	Layered	38.9	0.177	0.284	475	2.45	4.56
	Mixed	33.3	0.166	0.265	465	2.16	3.45
Comp	Layered	32.7	0.149	0.263	479	1.96	3.10
	Mixed	31.9	0.166	0.266	368	1.90	2.82
Comp+	Layered	31.2	0.143	0.260	387	3.90	3.56
Chisel	Mixed	30.1	0.140	0.236	473	2.29	2.94
LSD (p=0.05)		NS[b]	NS	NS	NS	NS	NS
LSD (p=0.10)		NS	NS	NS	NS	1.73	NS

[a] Not = not compacted; Comp = Compacted; Comp+chisel = Compacted plus chisel
plow.
[b] NS = not significant.

occurred when the soil was likely frozen (Table III) and runoff was therefore not influenced by the soil differences in bulk density and infiltration rates (Table II).

As was typical throughout this study, the quantities of N leached were more than 20 fold greater than quantities of runoff N (Table V). Even so, the amount of N leached equated to only 1.7% of the fertilizer N applied. Soil treatment effects on leachate volume and N leached were not significant. There was a non-significant trend of less N leaching when the topsoil-subsoil interface was disrupted by tillage rather than simply layering topsoil over the subsoil.

The possibility that dominance of winter runoff masked potential influences of the soil disturbance treatments prompted examination of runoff data collected during project year 1 and 2 growing seasons. Means of runoff for the 2 years and N and P losses failed to disclose any significant soil treatment effects (Table VI).

Fertilization and Clipping Management

The transition from soil manipulation treatments in project years 1 and 2 to fertilizer and clipping management treatments in years 3 and 4 was made with the assumption that, due to lack of significant treatment effects in the first two years (Table IV), there would be no confounding of the results obtained in years 3 and 4. To test this assumption, data from summer, winter, and all of year 3 were regressed on comparable data from year 2. None of the resulting R^2 values were significant (p = 0.108–0.617), indicating that the soil manipulation treatments in years 1 and 2 did not influence the results observed in years 3 and 4 for the fertilizer-clipping management treatments.

Table VI. Means of Project Years 1 and 2 Growing Season Runoff Water and N and P Loss as a Result of Soil Disturbance Prior to Lawn Establishment

Soil Disturbance[a]		Depth	Runoff Loss N	P
Subsoil	Topsoil	mm	------------ kg ha^{-1} ------------	
Not	Layered	8.8	0.103	0.051
Compacted	Mixed	11.8	0.092	0.047
Compacted	Layered	9.1	0.086	0.031
	Mixed	9.8	0.080	0.040
Compacted	Layered	10.4	0.094	0.036
+Chisel	Mixed	12.8	0.086	0.034
LSD (p=0.05)		NS[b]	NS	NS
LSD (p=0.10)		NS	NS	NS

[a] Not = not compacted; Comp = Compacted; Comp+chisel = Compacted plus chisel plow.
[b] Not significant.

Means of data collected over 2 years (Table VII) or for individual years (data not shown) failed to substantiate claims that natural organic fertilizers are less harmful to the environment than synthetic fertilizers. Quantities of runoff and N and P loads were not significantly different when comparing the two treatments. This was also true for leachate volumes and leachate N.

Lack of fertilization for 2 years increased the amount of runoff by 31 to 38% and altered runoff N and P losses (Table V). Phosphorus losses were most affected, showing increases of 50 to 58% above those of the fertilized plots. Not fertilizing increased runoff N by 25% when plots were mulch mowed, but had no effect when clippings were removed. These consequences of not fertilizing lawns have been observed elsewhere (19) and are attributable to the fact that regular fertilizer N applications are necessary to maintain high turfgrass stand density (20). High stand densities provide greater resistance to surface water flow, high residence times, more infiltration, and less runoff (19, 21). With high turfgrass stand densities, runoff water P loads do not increase even when compared to fertilized turfgrass where substantial amounts of fertilizer P are annually applied (19).

When averaged over the fertilization treatments, mulch mowing reduced the depth of runoff water by 9.1%, runoff N by 12.8%, and runoff P by 12.7% as compared to clipping removal (Table VII). These reductions were not statistically significant. Clipping disposition did significantly influence leaching losses of N. For the fertilized plots, nitrate-N concentrations increased by 65% and leaching N loss by 48% as a result of leaving the clippings on the plots. Clipping disposition did not affect leaching of N when no fertilizer was applied,

Table VII. Mean Annual Runoff Water and N and P loss as Influenced by Lawn Fertilization and Grass Clipping Disposition

Fertilizer Applied	Clipping Disposition	Runoff Loss			Leachate Loss		
		Depth mm	N ---- kg ha^{-1}	P ----	Depth mm	NO$_3$-N mg L^{-1}	N kg ha^{-1}
Synthetic	Removed	37.5	0.255	0.387	327	4.82	3.05
29-3-4	Mulched	34.9	0.205	0.296	350	8.36	4.65
Biosolids	Removed	40.8	0.279	0.375	356	5.62	3.27
6-2-0	Mulched	38.4	0.232	0.355	389	8.88	4.74
No	Removed	54.2	0.279	0.573	394	4.62	4.41
Fertilizer	Mulched	48.1	0.272	0.514	414	4.71	4.47
LSD (p=0.05)		6.6	NS[a]	0.169	NS	4.23	NS
LSD (p=0.10)		5.4	0.062	0.138	NS	3.46	1.62

[a] NS = not significant.

possibly because of less biomass and lower N concentration in clippings from the unfertilized treatment. There were instances when in the unfertilized plots almost no grass regrowth took place between mowings.

Nitrogen Carrier Effects

Treatment variables during the last two years of the study were fertilizer N carrier and P rates (Table VIII). The fertilizer treatments employed in project years 3 and 4 had the potential for confounding the effects of N carriers on runoff and N and P losses in years 5 and 6. To partially offset this, in October of year 4, the plots not fertilized in years 3 and 4 received an application of 98 kg N ha[-1] with the goal of restoring turfgrass density on those plots to a density comparable to that of the other plots. Linear regression of N loss in the summer of year 5 on summer N loss in year 4 indicated that the fertility treatments in years 3 and 5 significantly influenced summer N loss in year 5 ($R^2 = 0.401$, p = 0.0064). Results of similar regression analyses for winter and annual N losses and for runoff depths and P losses for summer, winter and the entire year gave no indication that year 3 and 4 treatments confounded treatment effects during these time frames in year 5 and presumably in year 6 as well.

The 2 year means for quantities of runoff water, runoff N and P, leachate and N leached were greater than for any other period in the study. The primary reasons were the quantities and distributions of precipitation in the second year (project year 6). Unusually large amounts of rainfall in June and July and 118 mm of precipitation preceding snowmelt in March (Table I) decidedly increased runoff.

The large increases in runoff amplified variability in N carrier treatment effects on N and P losses, and there were no significant differences (Table VIII). The initial thought was that the treatment randomization process inadvertently concentrated certain treatments on plots that had received no fertilizer for the previous two years. This proved not to be the case. Within the three blocks of treatments, only once was any single N carrier treatment assigned to a previously unfertilized plot. The true causative factors for what appeared to be significant N carrier influences on runoff water volumes and N and P loads were not readily apparent. A possible factor will be discussed in the Summary section.

There was a low degree of correspondence between runoff water depths and the quantities of N lost (Table VIII). In fact, the greatest N loss was in the biosolids-treated plots where runoff volume averaged 20 mm less than from the urea, PSCU, and PCU plots. What this indicates is that N loss per unit of runoff was unusually high when the biosolids fertilizer was applied and considerably higher than for the majority of the synthetic fertilizers. The data of Easton and Petrovic (19) show the opposite to be true when comparing biosolids with controlled-release fertilizers.

For the three treatments where P was applied, soluble runoff P averaged 0.98 kg ha[-1] (Table VIII). Not applying P reduced the amount of runoff P by

Table VIII. Annual Runoff Water and N and P Losses as Influenced
by Fertilizer N Carrier and P Applied

N Carrier[a]	P Rate $kg\ ha^{-1} yr^{-1}$	Runoff Loss			Leachate Loss		
		Depth mm	N	P	Depth mm	NO_3-N $mg\ L^{-1}$	N $kg\ ha^{-1} yr^{-1}$
			-- $kg\ ha^{-1} yr^{-1}$ --				
Urea	0	73.2	1.23	1.29	627	2.79	4.85
Biosolids	28.6	53.0	1.57	0.99	662	2.76	5.46
PSCU	2.6	73.6	1.19	0.88	643	2.15	4.18
PCU	17.2	74.3	1.42	1.07	584	1.59	3.48
Meth.urea	0	44.6	0.90	0.67	616	2.36	4.66
IBDU	0	36.3	0.73	0.59	526	2.15	3.24
LSD (p=0.05)		27.9	NS[b]	0.45	NS	NS	NS
LSD (p=0.10)		22.8	0.65	0.36	230	NS	NS

[a] Biosolids = dried activated sewage sludge; PCSU = polymer + sulfur-coated urea; PCU = polymer coated urea;
Meth.urea = methylene urea; IBDU = isobutylidene diurea.
[b] NS = not significant.

13% to 0.85 kg ha^{-1}, a statistically insignificant amount. Although fertilizer P rates ranged from 2.6–28.6 kg ha^{-1}yr^{-1}, the relationship between runoff soluble P and fertilizer P rate was inconsistent and not significant (R^2 = 0.095). Taking into account treatment differences in runoff depth increased the R^2 to an insignificant value of 0.598 (p = 0.255).

The primary purpose in applying different N carriers was to observe carrier influences on leaching loss of N. No significant effects were detected for the mean annual nitrate-N concentrations or the quantities of N leached (Table VIII). The nitrate-N concentrations of 1.59–2.79 mg L^{-1} are comparable to results reported by other researchers (*1, 2, 19*) when differences in soil texture, N application rates, and soil depth and method of leachate collection are taken into account. Leaching losses of N averaged 4.3 kg ha^{-1}yr^{-1}, a quantity that is intermediate to quantities leached in comparable studies (*1, 2*). The amount of N leached when fertilizer N was not applied (Table VIII; *19*) was not significantly different from when N was applied and was generally far less than observed for field crops (*4, 22*).

Summary and Conclusions

High spatial variability is a common feature in investigations of runoff and runoff water nutrient loads (*10, 19*). The present study was no exception. One prominent source of spatial variability appeared to arise from lack of uniformity in snow cover. This was indicated by plot-by-plot winter runoff totals over five years (Figure 1). With winter runoff dominating annual runoff and N and P losses (Table III), the amount of variability arising in a given treatment and the means for runoff and N and P losses for that treatment were dependent on the plots to which the randomization process assigned a particular treatment.

Various methods can be used to manipulate data from runoff studies so as to reduce experimental error and increase the numbers of statistically significant treatment effects. One is to employ geometric rather than arithmetic treatment means (*23*). Non-homogeneous variances can be dealt with by appropriate data transformation (*19*). Bartlett's test of homogeneity of variances (*24*) indicated that this was not a problem in the present study.

The ultimate use of the information collected in this study is to identify those management strategies that will have the greatest impact on N and P losses from upper Midwest lawns. Inspection of data collected throughout this study (Tables V, VII, and VIII) leads to the observation that quantities of N and P in runoff water may have been more dependent on quantity of runoff water than variations in the N and P concentrations in the water. To verify this, runoff water N and P loads were regressed on runoff water depth. These regressions and their coefficients of determination (Figures 2 and 3) indicate that quantity of runoff water more than any other factor determined N and P losses. This was particularly true in the case for soluble P, where R^2 = 0.819. The inescapable

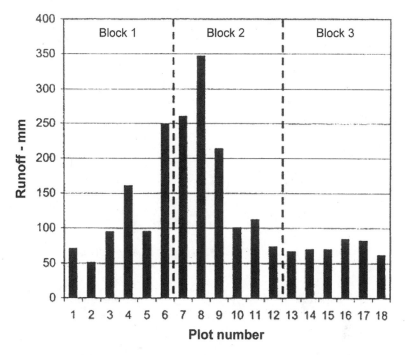

Figure 1. Five-year total runoff per research plot.

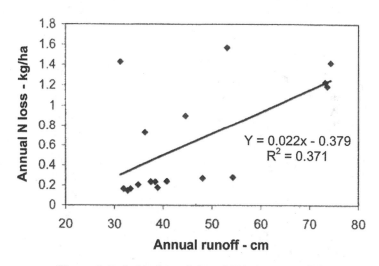

Figure 2. Relationship of annual N loss to runoff depth.

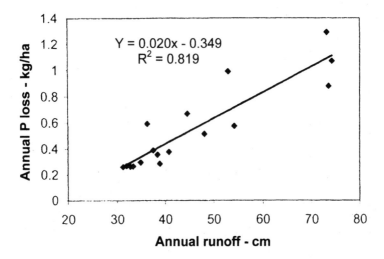

Figure 3. Relationship of annual P loss to runoff depth.

conclusion is that control of runoff is very likely the most important single step that can be taken to reduce soluble P runoff losses from upper Midwest lawns. The impact on runoff N loads is less certain.

In climatic zones such as the upper Midwest, where winter runoff may account for 68 to 99% of annual totals (Table III), there are few options for reduction of N and P losses from lawns. Maintenance of sufficient turfgrass density to entrap and prevent sediment loss and increase growing season infiltration is perhaps the best option. Increasing water infiltration rate via chisel plowing of compacted subsoil may have some value, but appears to be no more effective than tilling topsoil into the subsoil (Table V).

This study failed to provide evidence that not applying fertilizer P when soil test P was excessive will significantly reduce runoff P losses when no sediment loss occurs (Table VIII). Furthermore, the data of Barten and Jahnke (26) show that even when sediment losses occur, runoff water total and soluble reactive P concentrations bear no significant relationships to soil test P levels. Therefore, use of soil test P as criteria for regulating fertilizer P application is inappropriate.

Regulatory agencies need to be made aware that fertilizer and soil are not the only sources of P in lawn runoff water. A prominent source is vegetation in the landscape (8, 9, 10); how prominent remains to be conclusively demonstrated by further research. Laboratory measurements of P leached with water from grass clippings taken from research plots in this study (Table VIII) indicated that the grass itself could potentially have been the sole source of P in both summer and winter runoff water. The obvious implication is that as long as lawns remain a prominent feature in urban landscapes, much of the P in runoff water is not amenable to regulation.

Indications from this study (Tables V, VI, and VIII) and others (*2, 27, 28*) are that when lawns populated by cool season grasses are fertilized at N rates required to maintain high stand density, the quantities of N leached are not greatly different from unfertilized lawns, and leachate nitrate-N concentrations are not likely to pose a hazard with regard to groundwater drinking quality.

Precautions need be applied when extrapolating the results of the present study to other locales and lawns. These data are only applicable to climatic regions where annual runoff and N and P runoff losses occur predominantly during the winter months. The data only pertain to properly maintained lawns of sufficient density to prevent sediment loss. Lastly, the runoff P loads reported here are derived solely from measures of soluble P. Any additional forms of P that may have been present in the runoff water were not measured.

References

1. Gross, C. M.; Angle, J. S.; Welterlen, M. S. *J. Environ. Qual.* **1990**, *19*, 663-668.
2. Morton, T. G; Gold, A. J.; Sullivan, W. M. *J. Environ. Qual.* **1988**, *17*, 129-130.
3. Burwell, R. E.; Timmons, D. R.; Holt, R. F. *Soil Sci. Soc. Am. Proc.* **1975**, *39*, 523-528.
4. Randall, G. W.; Iragavarapu, T. K. *J. Environ. Qual.* **1995**, *24*, 360-366.
5. Dennis, J. *Proc. 5th Annual Conference of the North American Lake Management Society, Lake Geneva, WI.* **1985**, 401-407.
6. Singer, M. J.; Rust, R. H. *J. Environ. Qual.* **1975**, *4*, 307-311.
7. Timmons, D. R.; Holt, R. F. *J. Environ. Qual.* **1977**, 4, 369-373.
8. Timmons, D. R.; Holt, R. F.; Latterell, J. J. *Water Resour. Res.* **1977**, *6*(5), 1367-1378.
9. Dorney, J. R. *Water, Air and Soil Pollution* **1956**, *28*, 439-443.
10. Sharpley, A. N. *J. Environ. Qual.* **1981**, *2*, 160-165.
11. Brakensich, D. L.; Osborn, H. B.; Rowles, W. J. *USDA Handbook 224.* Washington, DC. 1979, pp 245-249, 375-377.
12. Holder, M.; Brown, K. W.; Thomes, J. C.; Zabrik, D.; Murray, H. E. *Soil Sci. Soc. Am. J.* **1991**, *55*, 1195-1202.
13. Bremner, J. M.; Keeney, D. R. *Anal. Chim. Acta* **1965**, *32*, 485-495.
14. Murphy, J.; Riley, J. P. *Anal. Chim. Acta* **1962**, *27*, 254-267.
15. Ruark, G. A. *Soil Sci. Soc. Am. J.* **1985**, *49*, 278-283.
16. Nelson, L. B.; Muckenhirn, R. J. *J. Am. Soc. Agron.* **1941**, *35*, 1028-1036.
17. Kelling, K.A.; Peterson, A. E. *Soil Sci. Soc. Am. Proc.* **1975**, *39*, 348-352.
18. Partsch, C. M.; Jarrett, A. R.; Watschke, T. L. *Trans. ASAE* **1993**, *36*, 1695-1701.
19. Easton, Z. M.; Petrovic, A. M. *J. Environ. Qual.* **2004**, *33*, 645-655.

18

20. Fry, J.; Huang, B. *Applied Turfgrass Science and Physiology;* Wiley: Hoboken, New Jersey, 2004; p 223.
21. Linde, D. T.; Watschke, T. L.; Jarrett, A.R.; Borger, J. A. *Agron. J.* **1995**, *87*, 176-182.
22. Logan, T. J.; Randall, G. W.; Timmons, D. R. *North Central Regional Research Pub. 268.* 1980, Ohio Agric. Res. Develop. Ctr., Wooster, OH.
23. Bannerman, R. T.; Owens, D. W.; Dodds, R. B.; Hornwe, N. J. *Water Sci. Tech.* **1993**, *28*, 241-259.
24. Snedecor, G. W.; Cochran, W. G. *Statistical Methods*; Iowa State Univ. Press, Ames, IA, 1967; pp 296-298.
25. Bresler, E. In *Modeling Plant and Soil Systems*; Hanks, R. J.; Ritchie, J. T., Ed.; Am. Soc. Agron., Madison, WI, 1991; pp 145-180.
26. Barten, J. M.; Jahnke, E. *Suburban Lawn Runoff Water Quality in the Twin Cities Metropolitan Area, 1996 and 1997*, Hennepin Reg. Park Dist., Maple Plain, MN, 1997.
27. Engelsjord, M. E.; Singh, B. R. *Can. J. Plant Sci.* **1997**, *77*, 433-444.
28. Miltner, E. D.; Branham, B. E.; Paul, E. A.; Rieke, P. L. *Crop Sci.* **1996**, *36*, 1427-1433.

Chapter 2

Determining Nitrogen Loading Rates Based on Land Use in an Urban Watershed

Zachary M. Easton[1] and A. Martin Petrovic[2]

Departments of [1]Biological and Environmental Engineering and [2]Horticulture, Cornell University, Ithaca, NY 14853

Due to the role of nutrients as limiting agents for eutrophication in fresh water and marine estuaries and as a human health risk in drinking water, greater scrutiny of land use in urban areas is needed. Few studies have attempted to determine the contribution of individual land uses to water quality degradation in urban areas. A 40% urban, 332 ha watershed in Ithaca, NY was selected as the research site. Runoff collected from 98 precipitation events over a two-year period and three land uses was analyzed for ammonium (NH_4^+-N), and nitrate (NO_3^--N) and mass losses were calculated. Monitored land uses included fertilized lawns (FL), urban barren (UB) areas and wooded (FR) areas. Stream gauges were installed at the stream entrance to the urban area and the watershed outlet to monitor the impact of the urban land uses on stream water quality. A multivariate analysis of the data revealed that the FL land use had significantly higher ($p<0.05$) nitrogen (N) loss than the other land uses in areas with shallow soil and high runoff potential. On the deeper soil where runoff was low, the FL land use had significantly lower ($p<0.05$) N loss than the UB or FR land uses. Precipitation derived N inputs had an influence on N lost from all land uses in the watershed, while N measured in throughfall under the FR land use canopy was of the same order of magnitude as the N lost in runoff from these areas. As the stream flowed through the urban area there was a significant increase ($p<0.05$) in stream flow rates under storm conditions and a significant decrease ($p<0.05$) in stream flow rates under dry conditions when compared to the undeveloped upper watershed. Areally weighted N loads in the stream were as high, and in many cases significantly higher ($p<0.05$), from the undeveloped upper watershed than following streamflow through the urban area. This indicated that the urban area may be a N sink. The results demonstrate that it is imperative to assess land use performance both spatially and under varying climatic conditions in order to reduce surface water contamination.

Introduction

As urban areas continue to grow in the majority of the United States, determining their impact on water quality is of the utmost importance. Concern about increasing pollution in urban waters has raised questions about the contribution of differing land uses to surface water contamination (1). While phosphorus (P) is the primary limiting agent in freshwater, N can be the limiting agent for eutrophication in many estuarine and marine systems (2) and can cause eutrophication, algal blooms, and impaired water quality at levels well below those considered safe for human consumption (e.g., <10 mg L^{-1} NO_3^--N). Nitrate concentrations as low as 1 mg L^{-1} can have a negative impact on fresh water (3), while levels of 0.1 mg L^{-1} NO_3^--N can cause problems in coastal waters (4).

Land use performance is increasingly important in mixed land use areas. The land use is expected to function as a filter and reservoir for drinking water, and to provide habitat and recreational benefits to residents. There is increasing scrutiny of how land uses impact water quality. In these mixed land use watersheds, there are numerous sources of contaminants, which can affect water quality. Many are anthropogenic, such as fertilizers and pesticides applied to home lawns, road deicing and traction enhancing materials (5) and some sources are natural, such as pollen from trees (6), leaching of nutrients from plant tissue (7, 8), or wet and dry atmospheric deposition (9). The impact of each source on pollutant levels in surface waters is heavily dependent on the characteristics of each watershed. When numerous contaminant sources are subjected to the high runoff losses inherent in developed areas, the impact on water quality can be great.

The increase in impervious areas associated with urbanization is a major cause of impaired water quality in many watersheds (10). Impervious areas can increase runoff volumes by presenting a physical barrier that prevents precipitation from infiltrating the soil, and in areas lacking storm sewers, increase the soil moisture levels in the surrounding soils and subsequently increase runoff and pollutant losses (11). Runoff in many storm sewered watersheds often discharges directly into streams and or surface water bodies, and short circuit the natural attenuation process provided by the soil. The Massachusetts Metropolitan District Commission, in charge of providing water to 40% of the state's residents, has identified urbanization as a major problem affecting water quality in the Quabin Reservoir Watershed (*12*). In one case, total N concentrations in urban storm water flows were measured in excess of 18 mg L^{-1} (*13*). The combination of these factors can cause N runoff losses from urban watersheds to be orders of magnitude higher than undeveloped watersheds (14).

While there is much speculation, and some evidence that urban areas are a source of N in surface water, there is also some evidence that residential developments have the potential to reduce N contamination. In a study in Baltimore, MD (*14*), NO_3^--N losses from agricultural watersheds (16.4 kg ha^{-1} yr^{-1}) were 3-5 times higher than urban (3.0 kg ha^{-1} yr^{-1}) or suburban watersheds (5.5 kg ha^{-1} yr^{-1}). The authors (*14*) identified residential developments as potential N sinks due to the significant amounts of lawn that may be present. Home lawns in Rhode Island were found to be a potential N sink (*15*). Over the two-year study, the average N mass losses were 1.35 kg ha^{-1} for both unfertilized lawns and forests (*9*). During the second, wetter year of the study the N mass losses from fertilized lawns were the same as the unfertilized and forested land uses. All three (fertilized, unfertilized, and forested land uses) had mass losses (1.5-1.9 kg ha^{-1}) an order of magnitude lower than from either agricultural land uses (41.8-100.0 kg ha^{-1}) or septic systems (47.5 kg ha^{-1}). These findings are intriguing if we consider that grassed areas are increasingly being considered as treatment sites for urban storm water (*16*). Therefore, practices that promote infiltration and subsequent uptake by plants can provide significant storage and biological remediation of non-point source pollutants (*17*).

This study was conducted to examine the contribution of three different land uses - fertilized lawns common to urban areas (FL), low or unmaintained urban land uses (UB) and wooded (FR) area - in a small urban watershed to stream N losses via runoff, as well as to determine the relative contribution of the various land uses.

Materials and Methods

Watershed Description

Located in Ithaca, New York (42°48'N, 76°46'W) in the Finger Lakes region, the experimental watershed is 332 ha in area (Figure 1), and is a sub-watershed in the Cayuga Lake basin in the Appalachian Plateau physiographic province. The region is typified by steep hillslopes with shallow permeable soils underlain by a restrictive layer of glacial origin.

Soils in the experimental watershed are generally deeper in the upper watershed (>100 cm) and underlain by bedrock. Soil depth decreases near the watershed outlet (<30 cm to 80 cm) and is underlain by fragipan, common to glacial tills. Soil particle distribution varies from a clay loam in areas near the watershed outlet to a fine sandy loam in the middle region of the watershed, and a silt loam in the upper area.. The climate is humid with an average annual temperature of 7.7°C and an average annual precipitation of 1143 mm. During the winter, soil freezing is common; thus soil temperatures were recorded at a depth of 100 mm, 3 km southeast of the watershed outlet. Elevations range from 250 to 350 m (Figure 1) with an average slope of 8.5%. The lower area of the watershed is urban (40% of watershed area) while the upper watershed is open water/wetland (8%), forested (44%), and pasture (8%) (Figure 1). The urban area is a mix of home land uses (lawns, wooded, ornamental, and impervious) as well as parkland, schools, and some light commercial development. Impervious surfaces (roofs, roadways, parking lots) were digitized and subsequently measured in a GIS, and comprise 24% of the lower watershed. There are very few storm sewers in the watershed, allowing the majority of the storm flows to drain via surface pathways. Those areas that are serviced by storm sewers discharge within the watershed. Septic systems and sanitary sewers serve a similar extent of the watershed

Land Use Selection and Plot Setup

Nine sampling locations in the urban area of the watershed were selected, consisting of three locations (3 replicates) for each land use type (Figure 1). Land uses were arranged in a block design on 7-11% slopes. Block I is located near the watershed outlet (Figure 1) on shallow, low storage soils, block II is in the middle of the watershed, and block III is at the top of the urban area on deeper, higher storage. Each plot is 2 m by 2 m, arranged parallel to the slope and bordered with steel edging inserted into the soil at a depth of 5 cm with 10 cm remaining above ground to prevent runoff from non-treatment areas from entering the experimental plot. This setup was shown in previous work (18, 19) to exclude surface runoff generated in non-treatment areas while allowing interflow (shallow subsurface flow) to enter the experimental plot. We allowed interflow to enter because this flow is the dominant runoff generating mechanism in the watershed (11). Runoff volumes were measured and collected at the soil surface with an aluminum angle (H flume) and directed into a tipping bucket system buried at a depth of 30 cm, collecting data continuously during runoff events (18). A subsample of each tip was collected and bulked for analysis.

23

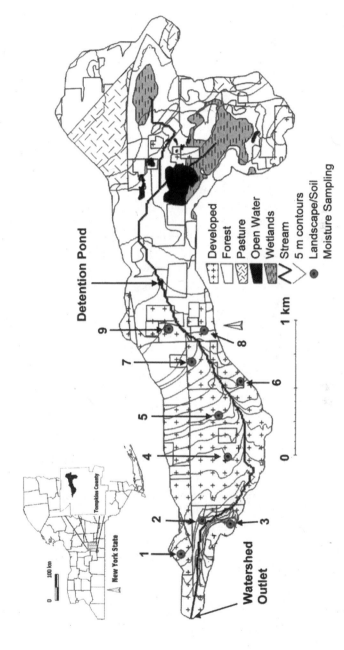

Figure 1. Watershed location in New York State (not to scale), and land use/land cover from the national land cover data center; locations of streams, land use runoff collection plots and stream gauges on 5 m contours.

On all UB land uses, we collected precipitation samples, recorded the volume and measured the NO_3^--N, and NH_4^+-N concentrations to determine the contribution from precipitation. Canopy throughfall samples were collected on all FR land uses, the volume recorded, and the NO_3^--N, and NH_4^+-N concentrations measured. The canopy throughfall measurements ultimately included the contribution of wet and dry N deposition and precipitation derived N. Previous work has shown that throughfall measurements are an viable method for estimating atmospheric deposition to forests and other eco-systems (20).

The fertilized lawn land use received fertilizer in three applications of approximately 50 kg N ha^{-1} (150 kg ha^{-1} y^{-1} N, 28 kg ha^{-1} y^{-1} P). While applications were made at approximately equal intervals during the growing season in May, August, and early November, the fertilizer sources and exact application timing varied between FL land uses. In 2003, fertilizer was applied to the FL land use in block 1 on 22-May, 14-August, and 3-November; and in 2004 on 18-May, 14-August, and 5-Novemeber. Fertilizer was applied to the block II FL plot in 2003 on 25-May, 22-August, and 7-November; and in 2004 on 1-June, 27-August, and 9-November. Fertilizer was applied to the block III FL plot in 2003 on 23-May, 14-August, and 2-November; and in 2004 on 29-May, 8-August, and 14-November. Unfortunately, application of the identical fertilizer treatments to the FL land uses was untenable, a result of the significant acreage and investment made by two of the three FL land use land owners, who account for approximately 25% of the fertilized land uses in the watershed. One FL land user made two applications of 25-5-10 (1.93% ammoniacal, 23.07% urea nitrogen (65% slow release poly coated urea, PCU), 10% P_2O_5 as $NH_4H_2PO_4$, 5% potash (K_2O) as muriate of potash), and one application in the fall of an 18-24-12 poly-coated urea (9.2% ammoniacal, 8.8% urea nitrogen (80% slow release PCU), 24% P_2O_5 as $NH_4H_2PO_4$, 12% potash (K_2O) as muriate of potash). Another FL land user made three applications of 25-8-8 (3.10% ammoniacal, 21.90% urea nitrogen (83% slow release PCU), 8% P_2O_5 as $NH_4H_2PO_4$, 8% potash (K_2O) as muriate of potash) The third FL land user applied two applications of a 33-4-11 (6.20% ammoniacal 26.80% sulfur coated urea (SCU) nitrogen, 4% P_2O_5 as $NH_4H_2PO_4$, 11% potash (K_2SO_4)). The third application was an 18-7-10 ((1.88% ammoniacal, 16.12% urea nitrogen (19% slow release SCU), 7% P_2O_5 as $NH_4H_2PO_4$, 10% potash (K_2O) as muriate of potash). The FL land uses are mowed weekly to 6 cm with clippings returned.

Kentucky bluegrass (*Poa pratensis* L.) and perennial ryegrass (*Lolium perenne* L.) are the predominant species in all three FL land uses, weed infestation is low, and plant density is high (100% ground cover). All FL land uses have been established for 10-15 years, and are generally of very high quality. The UB land uses receive no fertilizer, are mowed less frequently, if at all, and comprise 40-75% ground cover. One UB land use is a right of way for the power company, one is located at a private home, and one is located adjacent to a school. All three UB land uses are comprised of numerous species, including crabgrass (*Digitaria ssp.*), plantain (*Plantago ssp.*), ground ivy (*Glechoma* spp.), clover *(Trifolium, ssp.)*, wild blackberry (*Rubus, ssp*) and several small shrub species. The FR land uses receive no fertilization, and have little or no ground cover (0-55%). No definitive organic horizon exists on the FR land uses. All three contain predominantly deciduous tree species, including maple (*Acer spp.*), oak (*Quercus spp.*), beech (*Fagus spp.*), and ash (*Fraxinus spp.*), that are approximately 25-35 cm in diameter at breast height. None of the land uses are irrigated. Since groundcover for the UB and FR land uses was very low to none, it was quantified by visual estimate. For the FL land use, density counts were performed by placing a 9 cm^2 grid on the plot and counting all shoots with in the area. Shoot densities for the FL land use ranged from 8-10 shoots cm^{-2}.

Stream Gauges

Isco® model 6712 portable stream samplers with model 750 area velocity modules and model 674 tipping bucket rain gauges (Isco® Inc, Lincoln NE) were installed at the inlet to the urban area and the watershed outlet to record stream flow and rainfall on 10 minute intervals (Figure 1). The gauges at the inflow and outflow have two-part programs to collect base flow samples (i.e. when runoff is not occurring) and storm flow (runoff occurring). Flow weighted composite samples were collected every 1000 m^3 and bulked on a 5000-20000 m^3 basis, dependent upon overall flow dynamics. A stream distance of 2800 m separates the stream gauges.

Soil Sampling and Analysis

Prior to the initiation of the study, volumetric and gravimetric water content, porosity, bulk density, percent saturation, and moisture release characteristic curves were conducted on soil samples taken from each plot (three 44.75 cm^3 samples per plot) (*21, 22*). Soil particle density was determined by the pycnometer method (*23*). Soil porosity was corrected for organic matter content as determined by the loss on ignition method (*24*), and for gravel content (volume of gravel removed by a 2 mm sieve assuming a particle density of 2.44 g cm^{-3}). Bulk density ranged from 1.01 to 1.28 g cm^{-3}, particle density from 2.27 to 2.43 g cm^{-3}, porosity from 0.44 to 0.56 cm^3 cm^{-3}. Soil depth (soil surface to restricting layer) on the land use sampling plots was determined during installation (Fall 2002 – Spring 2003) of the runoff collection pit, by excavating the soil until the restricting layer was visible.

Runoff and Stream Water Sampling and Analysis

Land use runoff from 98 events (453 stream water samples), collected from April 2003 to April 2005, was stored at 4°C until analysis. Samples were generally analyzed within 10 days of collection. Samples were analyzed for nitrate+nitrite nitrogen ($NO_3^-+NO_2^-$-N, hereafter referred to as NO_3^--N), and ammonia nitrogen (NH_4^+-N). Nitrate and NH_4^+-N were analyzed at the Cornell Nutrient Analysis Laboratory (Cornell University, Ithaca, NY 14853). Nitrate was determined in water by auto-analyzer (hydrazine method) (*25, 26, 27*). Ammonium was determined in water by the automated phenate method (*28, 29, 30*).

Statistical Analysis

Land use and stream response variables were analyzed separately. Responses in each data set included runoff or stream flow depths, NO_3^--N, and NH_4^+-N mass losses. There were 98 runoff events in the land use data set and 453 water samples from the stream. Stream discharge data included 734 (daily) measurements. Land use location was used as the blocking variable for the land use data. A multiple analysis of variance (MANOVA) using a general liner model (GUB) was performed on natural log transformed response variables for both the land use and stream data in SAS (*31*). Type III sums of squares were used to determine the significance of the predictor variables in the regression analysis. A Fisher's Protected Least Significant Difference (FPLSD) was used to determine statistically significant differences between land use runoff volume, and nutrient mass losses at α=0.05. A Tukey's multiple comparison method was used to determine differences between flow and nutrient mass losses between stream gauges at α=0.05.

Results and Discussion

Runoff volumes and N mass losses varied significantly (*p*<0.05) between land uses, and blocks (Table I). However, the presence of a land use block interaction (Table I) precludes discussion of the main effects and forces one to discuss the results in terms of the blocking effect. The N input from precipitation was strongly correlated with N mass losses from all land uses, while throughfall N was significant only on the FR land uses during the growing season. The regression analysis revealed that while the concentration of N was significantly different for all land uses and stream gauging locations, the vast majority of the variation (75-96%) in the land use N mass losses or stream N loads regressions was due to differences in runoff volumes. Therefore, we will discuss only the runoff volumes and nutrient loads from the land use and in the stream.

Land Use Runoff

Surface runoff from the individual land uses was highly variable. Runoff volumes varied by a factor of 15 between land uses in different areas of the watershed, with the highest runoff volumes measured on shallower, finer-textured soil in block I (Figure 2). Surprisingly, the seasonal differences in runoff volumes were minimal, partially because both summers (generally the driest season in this region) during the study (2003 and 2004) were exceedingly wet (22 and 27% above normal, respectively). Major storms can occur at any time of the year. Seventy-eight percent of the summer runoff occurred during two events (spanning nine days total) in July 2003, and August/September 2004. Additionally, mid-winter snowmelt runoff in this region is generally a larger fraction of the total yearly runoff, but both winters were cold (average soil temperature at 100 mm depth was -2.5°C), and produced relatively small total runoff volumes. However, snowmelt in both 2004 and 2005 dominated the winter/spring runoff totals, with more that 82% of the total winter/spring runoff lost during snowmelt in March/April 2004, and 85% of total winter/spring runoff lost during snowmelt in March/April 2005. Average runoff volumes varied from nearly 6 mm event^{-1} in block I to less than 1 mm event^{-1} in block III (Figure 2).

The fertilized FL land uses had the lowest runoff volumes on the shallower, finer-textured soils in blocks I and II (Figure 2) compared to the UB and FR land

Table I. Significance Table for Multiple Analysis of Variance by Multiple Regression for Runoff Depth and N Mass Losses on ln Transformed Variables

Source[a]	Runoff	NH_4^+-N	NO_3^--N
		p-value	
Land use	<0.001	0.041	0.007
Block	<0.001	<0.001	<0.001
Season	0.341	0.018	0.007
Soil Depth	<0.001	0.043	0.012
Precipitation N Input	na	0.003	0.025
Throughfall N Input	na	0.763	0.667
Throughfall N x Land use x Season	na	0.041	0.032
Land use x Block	<0.001	0.037	0.014

[a] Source of variation is included in the model if significant at α ≤0.05. Main effects are included in the model if the interaction is significant at α ≤0.05.

uses. In block I, the UB land use had the highest runoff volumes, while in block II, the UB and FR land uses had the highest runoff volumes (Figure 2). Runoff volumes were similar between the land uses on the deeper, less runoff-prone soils in block III. Soil depth was the most significant predictor of land use runoff volumes in the watershed. Those land uses on shallow soil had significantly higher runoff volumes. The depth of the soil is important on these permeable soils, because runoff is created by soil saturation (11), and in most cases not by the infiltration capacity being exceeded. That is, the infiltration rate of these soils (4 - 27 cm hr^{-1}) is generally higher than the maximum intensity of the majority of precipitation events. When adequate precipitation falls, runoff occurs from the perched water table forming over the restricting layer in the soil and saturating the profile from the bottom up. Thus land uses situated on shallow soils or in areas with a large upslope contributing area are prone to larger, more frequent runoff losses due to the reduced storage capacity and the large volume of interflow from upslope. A simple regression analysis of land use runoff on soil depth yields a very strong negative relationship (Runoff = -0.09*Soil Depth + 8.91, $r^2 = 0.86$, $p<0.001$). This relationship indicates that a 10 cm increase in soil depth would result in an approximately 1 mm decrease in runoff which is important considering that the average runoff loss across land uses was 3 mm.

As indicated by the residual variation in the runoff soil depth regression, land use management can have an impact on runoff volumes as well. Runoff volumes from the FL land use on the shallow, runoff prone soils in block I fall below the regression line discussed above, indicating another possible source for the reduction in runoff volumes (Figure 2). Fertilization of the FL land uses may have reduced runoff volumes due to the high plant density (8-10 shoots cm^{-2} or 100% groundcover), assuming that fertilization increased plant density (18, 19, 31); whereas the UB and FR land uses had low plant density (0-45% ground cover). High plant density has been shown to dramatically reduce runoff volumes by creating a tortuous pathway, which reduces runoff velocities and increases infiltration, as well as reducing soil sealing by raindrop impact (32, 33). FL land use with densities of 10 shoots cm^{-2}, have been shown to have one third of the runoff volume and twice the infiltration rate of lawns with half the shoot density (19). Increased density has also been shown to reduce soil moisture levels by increasing evapotranspiration per unit area (34) and provides abundant organic matter at the soil surface which influences the movement of earthworms, and hence the creation of macropores capable of draining the profile (35). On the deeper sandier soils in block III, runoff volumes between the three land uses were similar (Figure 2).

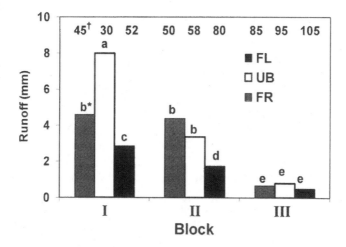

*Figure 2. Mean land use (fertilized lawn (FL), urban barren (UB), and wooded (FR)) runoff depths by block. *Land uses with the same letter are not significantly different as determined by a FPLSD at α≤0.05. † Approximate soil depth (cm) (surface to restricting layer) for land uses.*

Land Use Nitrogen Mass Loss in Runoff

Intuitively, landscape N mass losses were greatly influenced by the runoff volumes from the respective land uses. That is, high runoff volumes tended to result in high N mass losses (Figures 2 and 3). However, other factors may influence land use N loss as well, such as fertilization, N input from precipitation, or wet and dry N deposition. We considered N inputs from fertilizer, thought to be the largest source in urban watersheds (14), as well as from precipitation and wet and dry deposition. Ammonium mass losses were highest on the shallow, low storage soils in block I (0.346 kg ha[-1] event[-1]) and considerably lower on the deeper, high storage soils in block III (0.011 kg ha[-1] event[-1]) (Figure 3). The highest NH_4^+-N and NO_3^--N mass losses were measured on the FL and FR land uses on the shallow soils in block I and II, respectively. The lowest NH_4^+-N mass losses were on the deeper soils in block III and in block II on the FL land use (Figure 3). The lowest NO_3^--N losses were from the FL and UB land uses in block III and the FL land use in block II (Figure 3), correlating well with runoff volumes, fertilization, and soil depth. Higher runoff volumes, shallower soils, and fertilization of the FL land use in block I increased the mass loss of NH_4^+-N and NO_3^--N (Figure 3).

The high N mass losses from the FR land use in block II are mainly a result of high runoff volumes (Figure 2), but are also influenced by a higher available

N pool from organic matter turnover in the soil, and atmospheric N input from wet and dry deposition. We indirectly measured atmospheric N input by collecting canopy throughfall on the FR land uses, which would also include any N in rainfall, as well as any tissue N which could be mobilized. Strictly precipitation-derived N inputs were measured on the UB land uses. While the precipitation N inputs (both NH_4^+-N and NO_3^--N) were generally low (<0.1 kg ha^{-1}) the influence should not be ignored, as greater precipitation invariably resulted in greater N inputs (Figure 4). That being said, N inputs from precipitation were uniform across the land uses in this watershed, and would be expected to increase the available N pool on all land uses (e.g., no interaction effect with land use was detected, $p<0.05$ for both NH_4^+-N and NO_3^--N (Table I).

Compared with precipitation, higher N inputs were measured in canopy throughfall on the FR plots, but only during the growing season (May – November), when leaves were present ($p<0.05$, Table I, Figure 5). Throughfall N input during the growing season was significantly higher ($p<0.05$) than precipitation-derived N inputs (i.e. compare Figures 4 and 5). During the dormant season, there was not significant difference ($p<0.05$) between precipitation derived N input and canopy throughfall derived N inputs, not surprising, since there was no canopy in the deciduously forested areas. Similar to the precipitation N input, there was a strong correlation between throughfall N inputs and precipitation (Figure 5). However, it was non-linear during the growing season, with N inputs from throughfall reaching an asymptote after approximately 30 mm of precipitation (Figure 5), indicating that available N had been washed out of the canopy. During the dormant season, throughfall N inputs were linearly related to precipitation (Figure 5), similar to precipitation N input (Figure 4). While these N inputs were not extremely high (maximum 0.2 kg ha^{-1}, average 0.110 kg ha^{-1} during the growing season), they are of the same order of magnitude as the N losses from the FR land uses (Figure 3).

The FR land use in block I did not have a visible accumulation of leaf litter or organic matter present, so despite the high runoff losses (Figure 2) there may not have been much available N for transport via runoff. Microbe and earthworm activity in the soil is greatest during the summer, resulting in increased mineralization of organic matter, and subsequent conversion to NH_4^+-N in the soil and decomposing leaf litter layer (36, 37). Higher earthworm populations have been reported in urbanized forest soils, which increase N mineralization rates over natural forests soils (38). Earthworms are capable of depositing substantial ammoniacal N in casts, as high as 23 µg N g earthworm^{-1} d^{-1} (39).

The variability in N mass losses from the land uses in block I cannot be entirely attributed to differences in runoff volumes, as N mass losses were significantly higher from the FL land use, despite much lower runoff volumes (Figure 3) This is where the fertilization effect becomes clear. In this study the fertilized FL land uses showed increased plant density (8-10 shoots cm^{-2}) compared to the UB and FR land uses, which reduced runoff volumes, but due to

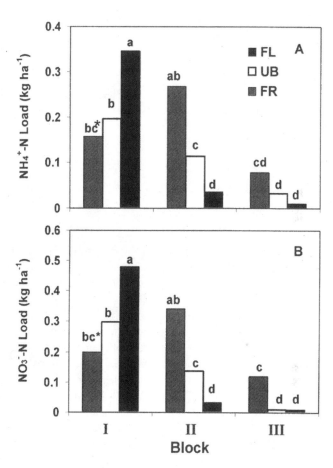

*Figure 3. Mean land use (fertilized lawn (FL), urban barren (UB), and wooded (FR)) ammonium (A) and nitrate (B) losses by block. *Land uses with the same letter are not significantly different as determined by a FPLSD at α≤0.05.*

32

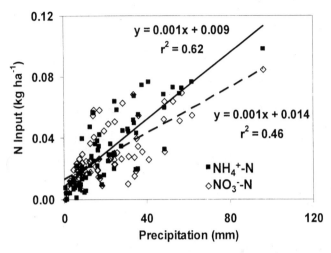

Figure 4. Relationship between precipitation depth and nitrogen (NH_4^+-N and NO_3^--N) input to the watershed from precipitation.

the highly runoff prone characteristics of the block I soils, and the increase in available N from soluble fertilizers, N mass losses were increased. The relative scale of the N losses from the land uses in blocks II and III were similar to their respective runoff volumes (Figures. 2 and 3).

Mass losses of NH_4^+-N were highest in the Spring and Summer (Table II), which is to be expected, given the increased mineralization rates (*36, 37, 38*) and generally high runoff volumes in Spring and Summer months during this study. The NH_4^+-N losses in the Spring and Summer were more than twice as high as the Fall, and more than an order of magnitude higher than those measured in the Winter. This may indicate that NH_4^+-N was rapidly converted to NO_3^--N in the soil during the Fall (*40*) and reduced the activity of mineralizing organisms as the soil cooled (*36, 37*). The highest NO_3^--N mass losses were found in the Spring, followed by the Summer and Fall (Table II), reflecting the influence of fertilization, N mineralization, and high runoff volumes. The high NO_3^--N losses in the Spring are a result of an increasing N pool from mineralization of organic matter on the FR and UB land uses, and Spring fertilizer application on the FL land uses (*41*) in conjunction with higher runoff volumes on more saturated soils. Other plant factors that were not measured may have also contributed to high NO_3^--N losses in the Spring.

The total NO_3^--N losses over the two study years were 10.12, 7.76, and 20.42 kg ha^{-1} from the FL, FR, and UB land uses respectively. Total NH_4^+-N losses were 7.75, 6.27, and 12.68 kg ha^{-1} from the FL, FR, and UB land uses, respectively. The UB land use had higher total N loading rates due to

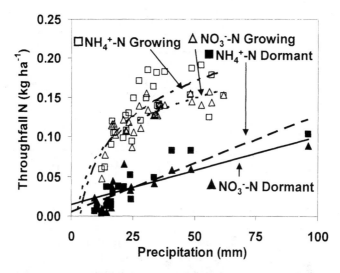

Figure 5. Canopy throughfall N by season (growing or dormant) measured on the FR land uses. Regressions are all significant at p<0.05. Equations are: NH₄⁺-N growing, y=0.06Ln(x)-0.08, r²=0.59*; NO₃⁻-N growing, y=0.05Ln(x)-0.05, r²=0.63**; NH₄⁺-N dormant, y=0.001x+0.005, r²=0.61*; and NO₃⁻-N dormant, y=0.001x+0.014, r²=0.65**. *indicates significance to the p<0.05 level.

significantly ($p<0.05$) more runoff events than either the FR or FL land uses. The total NO_3^--N losses from block I, two, and three were 79.47, 52.35, and 6.31 kg ha^{-1}, respectively. Total NH_4^+-N losses were 54.15, 27.19, and 5.57 kg ha^{-1} from blocks I, II, and III respectively. Block I had significantly higher ($p<0.05$) N loading rates due to more runoff per event and more events than blocks II and III.

Streamflow

The normalized streamflow (liters/m^2) in the watershed was highly variable, with flows ranging from 0.30 to 23.91 mm d^{-1} at the watershed outlet and from 0.37 to 13.72 mm d^{-1} at the inflow to the urban area (Figure 6). The normalized average daily discharge at the watershed outlet over the course of the study was 1.76 mm, while at the inflow to the urban area, discharge was 1.64 mm, not statistically different ($p=0.361$). However, discharges under different flow conditions are statistically different. During storm events the normalized discharge at the outlet was 3.44 mm, while at the inflow it was 2.78 mm, significantly lower ($p=0.030$). Conversely, under base flow conditions, the

Table II. Seasonal Average Land Use Mass Losses of Grouped Ammonium Nitrogen (NH_4^+-N), and Nitrate Nitrogen (NO_3^--N) Mass Losses in Runoff

Season	n	NH_4^+-N	NO_3^--N
		——————kg ha^{-1}——————	
Spring	85	0.204 a[a]	0.143 a
Summer	158	0.289 a	0.104 b
Fall	151	0.104 b	0.084 b
Winter	124	0.018 c	0.042 c

[a] Treatments are significantly different if the column means are not followed by the same letter as calculated by a Fisher's protected LSD at $\alpha \leq 0.05$.

normalized discharge at the outlet is 0.52 mm, while at the inlet it is 0.67 mm, significantly higher ($p=0.042$). This indicates that the two areas of the watershed responded differently under storm and baseflow conditions.

Stream response to precipitation was rapid and much greater in magnitude in the urban area of the watershed than from the undeveloped upper region (Figure 6). The impervious surfaces (roads, roofs, parking lots, etc.) intercepted precipitation before it could enter the soil and in many areas channelled it directly into the stream. In other areas, runoff from the impervious surfaces increased the soil moisture levels in the surrounding soil, and caused runoff to occur frequently, rapidly, and of a greater magnitude than in the upper undeveloped area (11). Runoff from the impervious surfaces, coupled with the shallow, low storage soils in block I tended to further increase runoff volumes near the watershed outlet, causing a rapid increase in runoff to the stream and a rapid rising limb to the stream hydrograph (42). The receding limb of the hydrograph was rapid as well, due to runoff production on the impervious surfaces ceasing immediately following the cessation of the precipitation. Conversely at the inflow to the urban area, both the rising and receding limb of the stream hydrograph were much shallower due to greater storage capacity of the soils in the upper undeveloped area of the watershed. This caused the stream discharge under runoff conditions to be 16% lower, with a peak discharge difference of 48% from the watershed outlet (Figure 6). Others have found urban areas to increase stream flow as well (43).

Under baseflow conditions, the normalized discharge at the watershed outlet was 22% lower than the normalized discharge at the inflow to the urban area. Thus, the stream was actually losing water in the urban area. This would indicate the water table feeding the stream during wet conditions had become a sink instead of a source of water. The higher runoff volumes in the urban area would lead to less water infiltrating the soil, and subsequently less recharge of the stream under base flow conditions. This is concerning for a number of reasons, because receiving waters, such as lakes or estuaries depend greatly on consistent

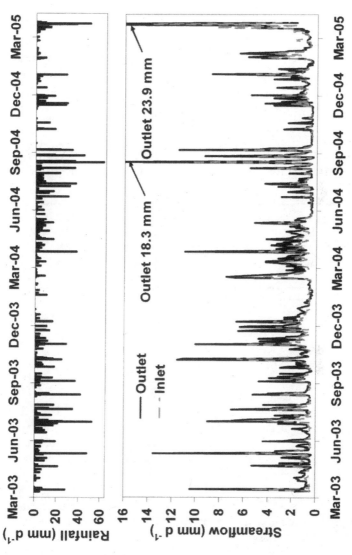

Figure 6. Inlet and outlet normalized streamflow and precipitation over the 2 year course of the study.

sources of water to maintain ecosystem function. Streams that have high flow rates under storm conditions but low flow rates during dry periods are likely delivering contaminant laden runoff to receiving waters (*44*), while delivering less water under base flow conditions that might serve to dilute the contaminants from runoff. Regardless, it is clear that the urban area in this watershed has a significant influence on the dynamics of stream flow.

The upper watershed, dominated by forest and wetland had a large storage capacity capable of infiltrating precipitation during storm events and slowly releasing the water following the precipitation event, thus supplying steady base flow conditions, and causing a lag in the stream response to precipitation from the upper watershed, which reduced the stream discharge during runoff events compared to the urban area.

Table III. Significance Table for Multiple Analysis of Variance by Multiple Regression for Streamflow and Stream N loads on ln Transformed Variables

Source[a]	Flow	NH_4^+-N	NO_3^--N
		— p-value —	
Sampling Location	0.001	0.502	0.001
Discharge	na	<0.001	<0.001
Location x Discharge	na	<0.001	<0.001

[a] Source of variation is included in the model if significant at $\alpha \leq 0.05$.
Main effects are included in the model if the interaction is significant at $\alpha \leq 0.05$.

Stream Nitrogen Load

Despite a significant increase ($p=0.001$) in stream flow during storm events attributed to the urban area of the watershed, the normalized NH_4^+-N load was significantly higher at the inlet to the urban area than it was at the watershed outlet (Figure 7A). In fact, during numerous large precipitation events, the normalized NH_4^+-N load was three to six times greater at the inflow to the urban area than the watershed outlet (Figure 7A). Normalized NH_4^+-N loads at the watershed outlet were generally less than 0.02 kg ha^{-1} d^{-1}, whereas the load at the inflow was as high as 0.05 kg ha^{-1} d^{-1}. This indicates that the urban area is acting as a sink of N, which may not be surprising due to the considerable acreage of high N demanding species, such as the lawns present in the urban area (*14*). The high NH_4^+-N loads from the upper undeveloped area of the watershed may be due to the mineralization of organic matter on the forest floor. Furthermore, mineralization from microbes and earthworms was abundant in the forested area creating a greater pool of mobile N (*38*). Additionally, waterlogged soil in the upper undeveloped area and a higher pH of the stream water at the inflow to the urban area (average pH of 6.3 and 5.5 at the inflow and watershed outlets, respectively tested on 7 dates), would increase the availability of the NH_4^+-N species over NO_3^--N. Water logged (anaerobic) soils favor the conversion of NO_3^--N to gaseous N_2 (a more favorable reaction), further reducing the availability of NO_3^--N in the soil.

Normalized NO_3^--N loads were approximately equal at the inflow to the urban area and the watershed outlet, and moved in unison (Figure 7B). Nitrate loads in the stream at both gauging locations were significantly higher ($p<0.05$) than the NH_4^+-N loads (Figures 7A and B). The largest NO_3^--N loads in the stream were during the Summer, approaching 0.12 kg ha^{-1} d^{-1}. This may reflect numerous natural processes such as greater mineralization and nitrification rates, heavy summer precipitation, or anthropogenic effects, such as fertilizer additions to the urban area. The high NO_3^--N load at the watershed outlet compared to NH_4^+-N (Figure 7B) was likely due to application of more soluble fertilizer based NO_3^--N in the urban area of the watershed.

Conclusions

The highest mass losses of N from all land uses were generally measured during the growing season on the shallow, low storage soils where runoff was more frequent and greater in magnitude. The UB land use in block I had the highest runoff volumes of any land use, due to shallow soil and little ground cover to slow runoff and allow infiltration. However, due to less available N in the UB land use, N mass losses were less than the fertilized FL land use. Interestingly, and counter to what might be expected, the NH_4^+-N and NO_3^--N mass losses were equally high from both the FL and FR land uses, despite fertilization of the FL land use. Nitrogen inputs to the FR land use from canopy throughfall (integrating the contribution of N from precipitation, wet and dry deposition) were the largest inputs to the FR land use, and of the same order of magnitude as measured N losses in runoff. The FL land use in blocks II and III (deeper, high storage soils) generally had significantly lower N mass losses and less runoff than either the UB or FR land uses. This indicates that controlling runoff is by far the most important avenue to realizing a reduction in nutrient contamination of surface waters.

Due to high runoff volumes in the urban area of the watershed, there was a dramatic increase in stream flow rates under storm conditions when compared to the predominately undeveloped forested area of the watershed. Conversely, under dry conditions, less water made it to the watershed outlet than flowed in at the inflow to the urban area. The low flow under dry conditions in the urban area was due to the impervious surfaces preventing rainfall from infiltrating, which increased runoff volumes and reduced groundwater recharge. In contrast, the

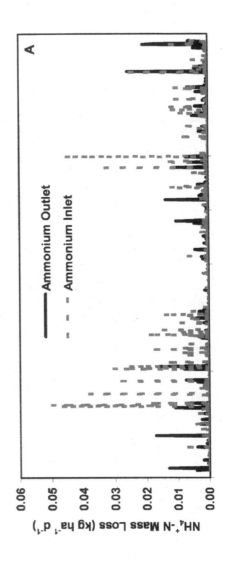

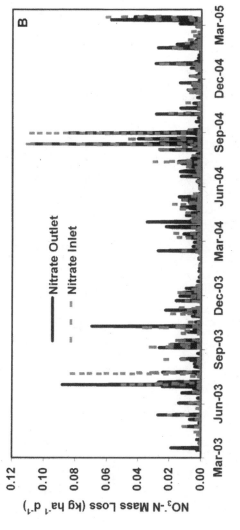

Figure 7. Normalized ammonium (A) and nitrate (B) loads in streamflow at the inlet to the urban area and watershed outlet. Loads are normalized on a kg ha^{-1} d^{-1} basis.

higher flow rates under storm conditions were a result of the greater area of impervious surfaces, and the shallower, finer textured soils creating more runoff in the urban area of the watershed. A reduction in storm flow by storm water management (infiltration basins, detention ponds) would be desirable to reduce N loss and the high storm flows which can cause stream bank erosion, and scouring. Furthermore, a reduction in storm flow would be the result of more runoff infiltrating the soil. This would subsequently increase flow during dry periods. Increasing infiltration would also allow the natural attenuation process to remove N through uptake, sorption to organic matter, or metabolization, further reducing N transport in the stream.

Nitrogen, particularly NH_4^+-N loads were as high and in many cases higher at the inflow to the urban area, than at the watershed outlet. In some cases the normalized (kg ha^{-1} d^{-1}) NH_4^+-N loads in the stream at the inflow to the urban area were six times higher than at the watershed outlet. Normalized NO_3^--N loads were approximately equal at the inflow and watershed outlet. The upper watershed, dominated by forested or wetland areas with significant organic matter accumulation, would be expected to be a source of nitrogen. However, it was not expected to be a larger source than the urban area, with fertilized lawns, and other areas such as impervious surfaces that have been shown to contribute N to the ecosystem. This finding supports the argument that urban developments may actually be a sink for N, due to the large N requirement of several land uses common to urban areas. However, based on these results it appears that there are specific land uses, management practices, and areas of a watershed that pose a higher risk to water quality degradation, such as the fertilized FL land use on the shallow runoff prone soils.

References

1. Interlandi, S.J.; C.S. Crockett. *Water Res.*, **2003**, *37*, 1737-1748.
2. Casey, R.E.; S.J. Klaine. *J. Environ. Qual.*, **2001**, *30*, 1720-1731.
3. Walker, W.J.; B.E. Branham. In *Golf Course Management and Construction: Environmental issues.* J. C. Balogh and W. J. Walker, eds; Lewis Publ., Chelsea, MI, 1992; pp 105-219.
4. Mallin, M.A.; T.L. Wheeler. *J. Environ. Qual.*, **2000**, *29*, 979-986.
5. Foster, G.D.; E.C. Roberts; B. Gruessner; D.J. Velinsky. *Appl. Geochem.*, **2000**, *15*, 901-915.
6. Banks, H.H.; J.E. Nighswander. *Symposium on Sustainable Management of Hemlock Ecosystems in Eastern North America,* 1999, GTR-NE-267, 168-174.
7. Tukey, H.B. *Ann. Rev. Plant Phys.*, **1970**, *21*, 305-324.
8. Sharpley A.N. *J. Environ. Qual.*, **1981**, *10*, 160-165.

9. Burian S.J.; G.E. Streit; T.N. McPherson; M.J. Brown; H.J. Turin. *Environ. Model. and Software.*, **2002**, *16*, 467-479.
10. Zandbergen, P.A. *J. Hazardous Materials*, **1998**, *61*, 163-173.
11. Easton, Z.M.; P. Gerard-Marchant; M. T. Walter; A. M. Petrovic; T. S. Steenhuis. *Water Resour. Res.*, **2007**, *43*, Article W03413.
12. Randhir, T.O.; R.O. Connor; P.R. Penner; D.W. Goodwin. *For. Ecol. Manag.*, **2001**, *143*, 47-56.
13. Brezonik, P.L.; T.H. StadeUBann. *Water Res.*, **2002**, *36*, 1743-1757.
14. Groffman, P.M.; N.L. Law; K.T. Belt; L.E. Band; G.T. Fisher. *Ecosystems*, **2004**, *7*, 393-403.
15. Gold, A.J.; W.R. DeRagon; W.M. Sullivan; J.L. Lemunyon. *J. Soil Water Conserv.*, **1990**, March-April, 305-310.
16. Deletic, A. *J. Hydrol.*, **2005**, *301*, 108-122.
17. Brown, W.; T. Schueler.. *National Pollution Removal Performance Database for Stormwater BMPs*. Center for Watershed Protection, Silver Spring, MD, 1997.
18. Easton, Z.M.; A.M. Petrovic; D.J. Lisk; I.M. Larsson-Kovach. *Int. Turfgrass Sci. Res. J.*, **2005**, *10*, 121-129.
19. Easton, Z.M.; A.M. Petrovic. *J. Environ. Qual.*, **2004**, *33*, 645-655.
20. Fenn, M.E.; M.A. Poth. *J. Environ. Qual.*, **2004**, *33*, 2007-2014.
21. Danielson, R.E.; P.L. Sutherland. In *Methods of Soil Analysis Part 1 Physical and Mineralogical Methods*. A. Klute, ed. ASA-CSSA-SSSA, Madison, WI, 1986; pp 443-461.
22. Gardner, W.H. In *Methods of Soil Analysis Part 1 Physical and Mineralogical Methods*. A. Klute, ed. ASA-CSSA-SSSA, Madison, WI, 1986; pp 493-544.
23. Flint, A.L.; L.E. Flint. In *Methods of Soil Analysis Part 4 Physical Methods Vol. 5*. J. H. Dane; G. C. Topp, eds ASA-CSSA-SSSA, Madison, WI. 2002, pp 229.
24. Nelson, D.W.; L.E. Sommers. In *Methods of Soil Analysis Part 3 Chemical Methods*. D. L. Sparks, ed. ASA-CSSA-SSSA, Madison, WI, 1996, pp 960-1010
25. USEPA. *Methods for Chemical Analysis of Water and Wastes. Method 351.1*. 1979.
26. Jacobs, M.B.; S. Hochheiser. *Anal. Chem.*, **1958**, *30*, 426-428.
27. Kamphake, L.J.; S.A. Hannah; J.M. Cohen. *Water Res.*, **1967**, *1*, 205.
28. USEPA. *Methods forChemical Analysis of Water and Wastes. Method 350.1*. 1979.
29. ASTM. *Standard D1426-79*. 1984, vol. 11.01:209.
30. Weatherburn, M.W. *Anal. Chem.*, **1967**, *39*, 971.
31. SAS Inst. Inc. *SAS/STAT User's Guide. Release V 9.1*. SAS Inst. Inc., Cary, NC, 2003.

32. Gross, C.M.; J.S. Angle; R.L. Hill; M.S. Welterlen. *J. Environ. Qual.*, **1991**, *20*, 604-607.
33. Krenitsky, E.C.; M.J. Carroll; R.L. Hill; J.M. Krouse. *Crop Sci.*, **1998**, *38*, 1042-1046.
34. Harrison, S.A.; T.L. Watschke; R.O. Mumma; A.R. Jarrett; G.W. Hamilton, Jr. In *Pesticides in Urban Environments: Fate and Significance, Vol. 522.* K. D. Racke; A. R. Lesile, eds. American Chemical Society, Washington, D.C., 1993, pp 191-207.
35. Hamilton, G.W., Jr.; D.V. Waddington. *J. Soil Water Conserv.*, **1999**, *3*, 564-568.
36. Whalen, J.K.; R.W. Parmelee. *Soil Ecol. App.*, **1999**, *13*, 199-208.
37. Groffman, P.M.; P.J. Bohlen; M.C. Fisk; T.J. Fahey. *Ecosystems*, **2004**, *7*, 45-54.
38. Steinberg, D.A.; R.V. Pouyat; R.W. Parmelee; P.M. Groffman. *Soil Bio. Biochem.*, **1997**, *29*, 427-430.
39. Krishnamoorthy, R.V. *Revue. d'Ecologie at Bilogie du Sol.*, **1985**, *22*, 463-472.
40. Jensen, M.B.; H.C.B. Hansen; N.E. Nielsen; J. Magid. *Acta. Agric. Scand. Sect.*, **1998**, *B.48*, 11-17.
41. Miltner, E.D.; Stahnke, G.K.; and Backman, P.A.. *Intern. Turf. Sci. Res. J.*, **2001**, *9*, 409-415.
42. Newman, B.D.; B.P. Wilcox; R.C. Graham. *Hydrol. Proc.*, **2004**, *18*, 1035-1042.
43. Hu, Q.; G.D. Willson; X. Chen; A. Akyuz. *Environ. Modeling Assessment*, **2005**, *10*, 9-19.
44. Pionke, H.B.; W.J. Gburek; A.N. Sharpley; R.R. Schnabel. *Water Resour. Res.*, **1996**, *32*, 1795-1804.

Chapter 3

Determining Phosphorus Loading Rates Based on Land Use in an Urban Watershed

Zachary M. Easton[1] and A. Martin Petrovic[2]

Departments of [1]Biological and Environmental Engineering and [2]Horticulture, Cornell University, Ithaca, NY 14853

Non-point phosphorus (P) loading to surface water can degrade water quality and impair habitat. As urban areas continue to grow in the United States, critical P source loading areas need to be identified to assess their impact on water quality. A 332 ha urban watershed in Ithaca, NY was selected and monitored for two years, with above average precipitation. Runoff collected from 98 precipitation events and three land uses was analyzed for dissolved P (DP), particulate P (PP), and total P (TP), with mass losses calculated. Monitored land uses included fertilized lawns (FL), urban barren (UB) areas and wooded (FR) areas. Stream gauges were installed at the stream entrance to the urban area and the watershed outlet to monitor the impact of the urban area on stream water quality. A multivariate analysis of the data revealed that the FL land use had higher DP losses than the other land uses on the shallow, low storage, runoff prone urban soils. If runoff volumes were low, the FL land use had similar or considerably lower DP losses. Particulate P mass losses were highest from the FR and UB land uses due to little or no ground cover to prevent erosive P losses. Total P losses from the land uses were highest on the shallow, low storage soils and surprisingly similar among land uses. As the stream flowed through the urban area, there was a two fold increase in P loads when compared to the undeveloped upper area of the watershed. This indicated that the urban area was contributing P to the stream. Best management practices in this and other similar watersheds should focus on reducing runoff and P loss from source areas to realize the largest reduction in P loading to surface water. Ultimately, it is imperative to assess land use performance under varying conditions in order to reduce P loads in surface water.

Introduction

Due to its role as a limiting nutrient for aquatic plant growth, phosphorus (P) is the primary concern in fresh water systems. Recent work done in the New York City Watershed indicates that P levels as low as 0.024 mg L^{-1} can cause the growth and subsequent proliferation of cyanobacteria (*1*). Unlike predominately agricultural watersheds, urban areas are comprised of a mosaic of land uses including both impervious surfaces and pervious land uses (*2*), coexisting at a fine spatial scale (*3*). The impact of each land use within an urban area can be expected to vary in a spatial-temporal manner and independently of each other (*4*). Therefore, it is unclear how, if, and when urban areas impact water quality. The function of these areas must be studied in greater depth, and much more intensively, to draw conclusions as to the role of urban areas in water quality and ecosystem function.

The contribution of urban areas to large scale water quality is generally unclear. However, some studies have shown that urban areas can be a considerable source of P. Phosphorus concentrations in urban stormwater flows have been measured in excess of 9 mg L^{-1} (*5*). The United States Geologic Survey (USGS) (*6, 7*) concluded that concentrations of TP were generally as high in urban streams as in agricultural streams, with more than 70 percent of sampled urban streams exceeding the USEPA desired goal (0.1 mg L^{-1} of TP) for preventing eutrophication. The density of development has been shown to influence P loading (*8*). Medium density residential sites (e.g., single family homes) had annual TP loadings of 0.6 kg ha^{-1}, while high density, urban sites had considerably higher TP loads of 1.1 kg ha^{-1} yr^{-1} - presumably due to a higher fraction of impervious surfaces (*8*). In two urban Wisconsin basins, lawns were estimated to contribute greater than 50% of the DP and TP loads to surface water (*9*). Wooded areas were also identified as potential P source areas. However, runoff was not flow weighted, nor sampled year round (*9*), making it difficult to draw conclusions about P loading rates.

This study was conducted to examine the contribution of three different land uses (fertilized lawns (FL), urban barren (UB), and wooded (FR) areas) in a small urban watershed to stream P losses via runoff. This watershed also presents a unique opportunity to compare P losses from two dichotomous land uses, urban and undeveloped, within the same watershed.

Materials and Methods

Watershed Description

Located in Ithaca, New York (42°48'N, 76°46'W), the experimental watershed is 332 ha in area and a subwatershed in the Cayuga Lake basin. The region is typified by steep hillslopes with shallow permeable soils underlain by a restrictive layer of glacial origin. Soils in the experimental watershed are generally deeper in the upper watershed (>100 cm) and underlain by bedrock, while soil depth decreases near the watershed outlet (<30 cm to 80 cm) and is underlain by a fragipan common to glacial tills. During the winter, soil freezing is common, thus soil temperatures were recorded at a depth of 100 mm, 3 km south east of the watershed outlet. Elevations range from 250 to 350 m, with an average slope of 8.5%. The lower watershed is urban (40% of watershed area) while the upper watershed is open water/wetland (8%), forest (44%), and pasture (8%). The urban area is a mix of home land uses (lawns, wooded, ornamental, and impervious) as well as parkland, schools, and some light commercial development. Impervious surfaces were digitized and measured in a GIS and comprise approximately 24% of the lower watershed. See the companion paper, "Determining Nitrogen Loading Rates Based on Land Use in an Urban Watershed"(*10,*) for a complete description of the watershed.

Land Use Selection and Plot Setup

Nine sampling locations within the urban area of the watershed were selected, consisting of three land uses in three locations (3 replicates). Land uses were arranged in a block design, on 7-11% slopes. Runoff volumes were measured and collected at the soil surface with an aluminum angle H flume and directed into a tipping bucket system buried at a depth of 30 cm, collecting data continuously during runoff events (*10, 11*). A subsample of each tip was collected for analysis.

Fertilized lawn land uses received fertilizer in three applications of approximately 50 kg N ha^{-1} (150 kg ha^{-1} y^{-1} N, 28 kg ha^{-1} y^{-1} P). While applications were made at approximately equal intervals during the growing season in May, August, and early November, the fertilizer sources and exact application timing vary between FL land uses. In 2003, fertilizer was applied to the FL land use in block I on 22-May, 14-August, and 3-November; and in 2004 on 18-May, 14-August, and 5-Novemeber. Fertilizer was applied to the block II FL plot in 2003 on 25-May, 22-August, and 7-November; and in 2004 on 1-June, 27-August, and 9-November. Fertilizer was applied to the block III FL plot in 2003 on 23-May, 14-August, and 2-November; and in 2004 on 29-May,

46

8-August, and 14-November. Unfortunately, application of the identical fertilizer treatments to the FL land uses was untenable, a result of the significant acreage and investment made by two of the three FL land use land owners, who account for approximately 25% of the fertilized land uses in the watershed. One FL land user made two applications of 25-5-10 (1.93% ammoniacal, 23.07% urea nitrogen, 10% P_2O_5 as $NH_4H_2PO_4$, 5% potash (K_2O) as muriate of potash), and one application in the Fall of an 18-24-12 poly-coated urea (9.2% ammoniacal, 8.8% urea nitrogen, 24% P_2O_5 as $NH_4H_2PO_4$, 12% potash (K_2O) as muriate of potash). Another FL land user made three applications of 25-8-8 (3.10% ammoniacal, 21.90% urea nitrogen, 8% P_2O_5 as $NH_4H_2PO_4$, 8% potash (K_2O) as muriate of potash). The third FL land user applied two applications of a 33-4-11 (6.20% ammoniacal 26.80% sulfur coated urea nitrogen, 4% P_2O_5 as $NH_4H_2PO_4$, 11% potash (K_2SO_4)). The third application was an 18-7-10 ((1.88% ammoniacal, 16.12% urea nitrogen, 7% P_2O_5 as $NH_4H_2PO_4$, 10% potash (K_2O) as muriate of potash).

The UB land uses receive no fertilizer, and are mowed less frequently, if at all. One UB land use is a right of way for the power company, one is located at a private home, and one is located adjacent to a school. All three UB land uses are comprised of numerous species, including crabgrass (*Digitaria ssp.*), plantain (*Plantago ssp.*), ground ivy (*Glechoma* spp.), clover *(Trifolium, ssp.)*, wild blackberry (*Rubus, ssp*) and several small shrub species.

The FR land uses receive no fertilization, and have little or no ground cover (0-55%). All three contain predominantly deciduous tree species, including maple (*Acer spp.*), oak (*Quercus spp.*), beech (*Fagus spp.*), and ash (*Fraxinus spp.*), that are approximately 25-35 cm in diameter at breast height. Since groundcover for the UB and FR land uses was very low to none, it was quantified by visual estimate. For the FL land use, density counts were performed by placing a 9 cm^2 grid on the plot and counting all shoots with in the area. See (*10*) for a complete description of watershed and land use characteristics.

Stream Gauges

See (*10*) for complete details of stream monitoring.

Soil Sampling and Analysis

Soil test phosphorus (STP) was determined in April 2003, August 2004, and April 2005 using the Morgan soil test extract (*12*). See (*10*) for complete details on soil analysis.

Runoff and Stream Water Sampling and Analysis

Land use runoff samples from 98 events and 453 stream water samples , collected from April 2003 to April 2005 were analyzed for dissolved P (DP), particulate P (PP), and total P (TP). Dissolved P is all of the P present in a sample filtered through a 0.45µm filter, and represents the bioavailable P, or that most prone to create eutrophic conditions in fresh water (13). Particulate P is P associated with soil particles, generally mobilized via erosive processes, and is generally less bioavailable. Total P is the measure of the total quantity of P in an unfiltered sample, and includes both DP and PP. Dissolved phosphorus samples were filtered through a 0.45-µm filter, and molybdate soluble reactive orthophosphate was determined colorimetrically by ascorbic acid method (14). Total P determination was performed on unfiltered samples by persulfate digestion (15), and subsequent colorimetric measurement by the ascorbic acid method described above. Particulate phosphorus (includes dissolved organic P) was determined as the difference between TP and DP.

Statistical Analysis

Land use and stream P losses were analyzed by a multiple analysis of variance (MANOVA) using a general linear model in SAS (16). A Fisher's Protected Least Significant Difference (FPLSD) was used to determine statistically significant differences between land use P mass losses at $\alpha=0.05$. A Tukey's multiple comparison method was used to determine differences between P loads between stream gauge locations at $\alpha=0.05$. For complete details on statistical methods, see (10).

Results and Discussion

Phosphorus losses varied significantly ($p<0.05$) between land uses and blocks (Table I). However, the presence of a land use block interaction precludes discussion of the main effects and forces one to discuss the results in terms of the block effect. Regression analysis revealed that while the con-centration of P was significantly different between land uses and stream gauging locations, the vast majority of the variation (88-93%) in the land use or stream P loss was due to differences in runoff or stream flow, respectively, and not concentrations. Therefore, we will discuss only the P loads from the land uses or in the stream. Runoff volumes are discussed in detail in (10) and shown in Table II. Soil depth in the watershed, as captured by the blocking scheme, was the most significant factor in predicting runoff volumes from the land uses. The soil depth gives an indication of the available storage capacity of a volume of soil. All else being equal, greater soil depth results in greater storage capacity and generally reduced runoff volumes in highly pervious soils. Shallow, low storage soils in block I had more than an order of magnitude more runoff from all land uses than the deeper soils in block III. See (*10*) for a detailed explanation of soil depth in relation to land use runoff volumes.

Table I. Significance Table for Multiple Analysis of Variance by Multiple Regression for P Mass Losses on ln Transformed Variables

Source[a]	DP	PP	TP
		p-value	
Land use	<0.001	0.002	NS
Block	0.002	0.001	0.003
Season	<0.001	0.001	<0.001
Land use x Block	0.002	0.009	0.019

[a] Source of variation is included in the model if significant at $\alpha \leq 0.05$.
Main effects are included in the model if the interaction is significant at $\alpha \leq 0.05$.

Land Use Phosphorus Loss

Land uses sited on deeper, high storage soils in the watershed (block III) had lower DP mass losses, while DP mass losses were highest from the shallow soils in block I (Figure 1). The major reason for the difference in DP mass losses across blocks was due to differences in runoff volumes. Runoff volumes were on average an order of magnitude higher on the shallow soils (block I) than from the deeper soils (block III) (Table II) (*10*).

Dissolved P losses ranged from a low of 0.001 kg ha^{-1} event^{-1} for the FL and FR land uses on deeper soils (block III) to 0.014 kg ha^{-1} event^{-1} for the FL land use on the shallow soils in block I (Figure 1). The high DP loss from FL land use in block I is a result of the high runoff volumes (2.90 mm event^{-1}) (*10*) in combination with the P fertilization of the land use. The FL land uses are fertilized with P (4.5 to 16.8 kg P ha^{-1} app^{-1}), which may be in excess of the P needed by the turfgrass and provides a readily available source of P when runoff occurs. In New York State, 2 mg P kg^{-1} soil (Morgan extract) was found to be adequate to supply P to the turfgrass plant (*17*). Therefore it is possible that application above this level may result in increased P loss via runoff, provided that the soil affinity for P has been satisfied (e.g., soil sorption capacity). Indeed, STP values ranged from 3.2 to 19.4 mg kg^{-1} on the Morgan scale, which is low to high STP. The highest STP values were on the FL land use and the lowest on the unfertilized UB and FR land uses. While fertilizer application timing was not controlled by the researchers, there were not any applications made within 48 hours of runoff-causing precipitation events. Because of the very low levels of runoff on the deep, high storage soil in block III (Table II) (*10*), and the few number of runoff events, the DP mass losses were not different among land uses (Figure 1).

Table II. Runoff Depths for each Block and Land Use in the Watershed

Land Use	Block I	Block II	Block III
		(mm)	
Urban Barren	7.76a[a]	3.82b	0.51e
High Maintenance	2.90b	1.74d	0.36e
Forested	4.53c	4.45b	0.26e

[a] Land uses with the same letter are not significantly different as determined by a FPLSD at $\alpha \leq 0.05$.

Particulate P (PP) represents the less bioavailable P fraction and is generally associated with soil particles and thus erosive processes. Particulate P mass losses were influenced by the land use block interaction as well. The FL land use had the lowest PP losses in every block (from <0.001 kg ha^{-1} event^{-1} in block III to 0.004 kg ha^{-1} in block I, Figure 1). The FR and UB land uses in block I had the highest PP loss (0.013 and 0.010 kg ha^{-1} event^{-1}, respectively, Figure 1). Much of the difference in the land use PP losses can be attributed to the level of ground cover, as the highest ground cover was found on the FL land use (8-10 shoots cm^{-2}), while the lowest ground cover was on the FR and UB land uses. The PP loss from the FL land use in block III was less than 0.001 kg ha^{-1} event^{-1}, reflecting complete ground cover. Dense turfgrass has been shown to reduce sediment loss via runoff (*18*) by the dense fibrous root system stabilizing the soil, as well as the canopy reducing the impact of raindrops on the soil surface.

On the shallow, runoff prone soils in block I, PP losses from the FL land use were 60 and 66% lower than PP losses from the FR and UB land uses, respectively. On the deeper, higher storage soils in blocks II and III, PP losses from the FL land use were between 90-100% lower that losses from the FR and UB land uses (Figure 1). Indeed there was very little sediment (data not shown) measured from the FL land use, a result of the dense ground cover.

As expected, TP mass losses were highest (p=0.003) on the shallow, low storage soils in block I where runoff was highest (Table II) and not significantly different among land uses (p=0.341) (Figure 1). The lowest TP losses were from the FL land use on the deeper soils in block III due to the low runoff volumes (*10*). While there was no land use effect on TP mass losses in block I (Figure 1), TP losses in blocks II and III were significantly different among land uses

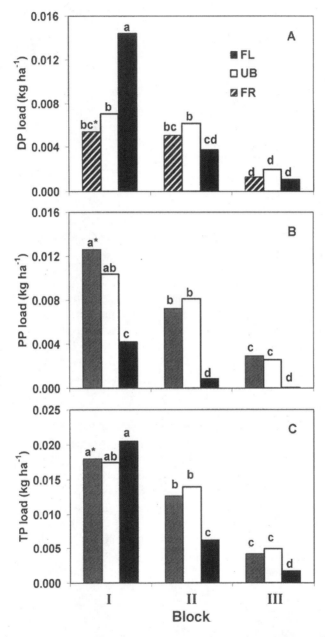

Figure 1. Mean land use (fertilized lawn (FL), urban barren (UB), and wooded (FR)) dissolved P (A), particulate P (B), and total P (C) losses by block for all 98 runoff events. *Land uses with the same letter are not significantly different as determined by a FPLSD at α≤0.05.

(Figure 1). In both blocks II and III, the TP mass losses from the FL land use were approximately half that of the UB and FR land uses, reflecting the lower runoff volumes in conjunction with lower DP and PP losses.

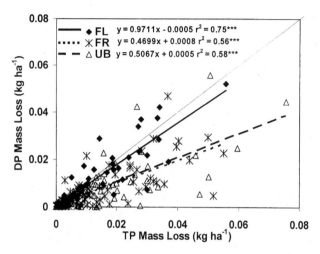

Figure 2. Relationship between total P (TP) and dissolved P (DP) mass losses by land use. *** indicates significance at the α= 0.001 level.

Figure 1 shows that the bulk of the P from the FR and UB land uses in block I was particulate, consistent with what would be expected with low ground cover land uses (*19*). The relationship did not appear to be consistent with the PP losses from the FR and UB land uses in blocks II and III and may be a result of the lower runoff volumes not having sufficient erosive power to remove sediment. The FR and UB land uses in block I had approximately 65% of the TP loss as PP (DP losses were relatively low). As would be expected, the FL land use had the vast majority of the TP lost as DP, consistent with fertilization (*21*), but the high DP fraction could also be from P leaching from plant tissue. However, comparing mean PP, DP, and TP losses directly with each other can be misleading, and the dynamics of the losses differ between P species. These differences in the mean values do not reflect differences in the temporal dynamics of P loss. For instance, intense thunderstorms in the summer can, in some cases, exceed the infiltration capacity of the soils and mobilize PP via erosion from the low ground cover FR and UB land uses. However, the low runoff volumes during the summer from the drier soils generally reduced DP losses. Therefore, it is more informative to compare ratios of the P species.

The relationship between DP and TP losses by land use (Figure 2) indicates that more of the TP is in the dissolved form from the FL land use. In fact, the DP losses from the FL land use very nearly fall on the theoretical 1:1 line with TP losses, again indicating minimal particulate contribution from the fertilized FL land use. The ratio of DP to TP mass loss was highest for the FL land use (76%) (Figure 2), reflecting the dense ground cover of the FL land use, which reduced the PP contribution, but also likely reflects the management influences of a high P application rate (relatively soluble). The DP losses from both the FR and UB land uses fall well below the 1:1 line with TP losses, which indicates a higher PP loss (Figure 2). This effect is due mainly to the PP losses from the block I FR and UB land uses; however, higher PP loads were measured from the FR and UB land uses in blocks II and III as well.

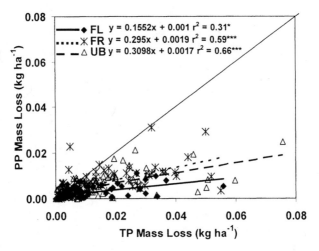

*Figure 3. Relationship between total P (TP) and particulate P (PP) mass losses by land use. *, *** indicate significance at the α=0.05 and 0.001 levels, respectively.*

The lower DP to TP ratio of the UB and FR land uses (48% each) indicates that PP was contributing comparatively more (52%) to TP losses. Figure 3 shows the relationship between PP and TP losses by land use. The slopes of the regressions for both the FR and UB land uses are nearly twice as steep as the slope of the FL regression, indicating that for FR and UB land uses there is considerably more PP lost. The relationship between PP and TP mass loss is also considerably stronger for the FR and UB land uses (p<0.001), than for the FL land use (p=0.045).

The highest DP losses independent of land use occurred in the Fall (0.016 kg ha^{-1}), nearly 36% of the total loss (Table 3). The high DP mass losses in the

Fall correlate well with P applications on the FL land use in November (16.5 kg P ha^{-1}) when the soil moisture level was higher and plant uptake was likely declining as temperatures approached freezing. High soil moisture levels (approaching field capacity) have been shown to increase available P levels (particularly dissolved forms) by causing anaerobic soil conditions and subsequent release of P from ferric hydroxides (20). Indeed, the soil was more consistently saturated in the Fall (22). Higher soil moisture levels also lead to higher runoff volumes, which invariably creates greater P loss, particularly for more soluble species such as DP. However, high variability in runoff volumes during the Summer and Fall reduced the statistical significance of the season by land use interaction. For the UB and FR, the higher runoff volumes almost entirely explain the higher Fall P losses. Particulate P losses were highest in the Summer (0.017 kg ha^{-1}), due to thunderstorms with intense precipitation removing more particulate associated P, notably from the low ground cover FR and UB land uses (Table III). Very little PP was found in runoff from the FL land use (Figure 1). Total P mass losses were highest in the Summer reflecting the higher contribution of PP from thunderstorms and in the Fall from the contribution of DP from fertilization. During the Winter when there is no fertilization and frozen conditions minimize the interaction of runoff with the soil, mass losses of all P species were the lowest (Table III).

Table III. Seasonal Average Dissolved Phosphorus (DP), Particulate Phosphorus (PP), and Total Phosphorus (TP) Losses in Runoff

Season	n	DP	PP	TP
			kg ha^{-1}	
Spring	85	0.011 b	0.008 b	0.018 b
Summer	158	0.013 b	0.017 a	0.025 a
Fall	151	0.016 a	0.008 b	0.024 a
Winter	124	0.005 c	0.003 c	0.008 c

NOTE: Treatments are significantly different if the column means are not followed by the same letter as calculated by a Fisher's protected LSD at α ≤0.05.

Total DP losses over the two study years were 0.53, 0.46, and 0.48 kg ha^{-1} from the FL, FR, and UB land uses, respectively, which is not significantly different. However, total losses for both PP and TP were significantly higher ($p<0.05$) from the UB and FR land uses. Total PP loss was significantly lower from the FL land use ($p=0.032$) (0.22 kg ha^{-1}) than from the FR, and UB land uses (0.67, and 0.57 kg ha^{-1}, respectively). Total TP loss was significantly lower from the FL land use ($p=0.043$) (0.77 kg ha^{-1}) than from the FR, and UB land uses (1.16, and 1.12 kg ha^{-1}, respectively). The total DP losses from blocks I, II, and III were 2.89, 2.01, and 0.74 kg ha^{-1}, respectively, which were significantly different (p=0.021). Total PP loss was significantly higher (p=0.029) from block I (2.90 kg ha^{-1}) than from blocks II and II (0.85, and 0.64 kg ha$^{-1,}$ respectively). Total TP loss was significantly higher (p=0.017) from block I (5.59 kg ha^{-1}) than from blocks II and II (2.81, and 1.25 kg ha^{-1}, respectively). Since there were fewer runoff events from the FL and FR land uses than the UB land use, and fewer events from blocks II and III, total two year losses were generally lower from these land uses or blocks (10).

The P losses we observed in this study fall with in the ranges reported by other researchers. In a study of residential sites (8), annual TP loading rates of 0.6 kg ha^{-1} were measured. A study in the Atlantic costal plain (23) measured P loss from a forested watershed of 0.26 kg ha^{-1} yr^{-1} as predominately particulate P, indicating that erosion is the pathway by which most P is lost from sites without dense ground cover. In Wisconsin (24), P fertilized lawns were estimated to contribute up to 0.70 kg ha^{-1} yr^{-1} of DP and 2.57 kg ha^{-1} yr^{-1} of TP to a lake from the 89 ha of lawns in the vicinity. Unfertilized lawns contributed 0.40 kg ha^{-1} yr^{-1} of DP and 1.73 kg ha^{-1} yr^{-1} of TP, and wooded areas contributed 1.04 kg ha^{-1} yr^{-1} of DP and 3.52 kg ha^{-1} yr^{-1} of TP (24). In another Wisconsin study (25), measured P losses from turfgrass were 0.42 to 0.58 kg ha^{-1} yr^{-1}. In New York (26), annual P loading rates in turfgrass runoff ranged between 0.2 and 1.3 kg ha^{-1} yr^{-1} depending on fertilizer source and P application rate. The highest P loss was from the low density-unfertilized turfgrass. A study in Pennsylvania (18) observed P losses in runoff ranging from 0.9 to 1.8 kg ha^{-1}yr^{-1}, where the lower P loss was from the denser bunch-type perennial ryegrass.

Other Sources of Phosphorus

Impervious Surfaces

Road runoff grab samples were taken periodically during the study (n=40) and analyzed for DP, PP, and TP. These samples were not flow weighted; therefore it is difficult to draw conclusions about P loads from impervious surfaces; however, an effort was made to sample at random times during a precipitation event, so as not to introduce bias by only sampling at the start (or end) of an event. As a result, some limited conclusions about the influence of P loading from impervious surfaces can be inferred.

Since the vast majority of precipitation falling on impervious surfaces is expected to end up as runoff, and this watershed has a very limited stormwater management system; it can be assumed that the bulk of the runoff is directed to the stream or areas adjacent to the impervious surfaces in the watershed. Concentrations of P in some samples of impervious surface runoff were excessively high (particularly during the Winter when traction and de-icing materials were spread on the roadways). On several occasions, TP concentrations in excess of 2 mg L^{-1} (2.66 mg L^{-1} highest) were measured. Granted, this is not a flow-weighted concentration, and these samples were taken at the onset of road runoff, but even limited runoff of this concentration can be detrimental. The mean TP concentration in the Winter was 1.20 mg L^{-1} and in the Summer it was 0.77 mg L^{-1}. However, it is likely that the concentrations of TP>2.0 mg L^{-1} were the result of the initial flush of P from the road. Samples taken at later times in road runoff events showed a dramatic reduction in P concentrations. In fact, if sampled after approximately the first 15 mm of precipitation, the road samples are relatively free of P (12 were below TP detection limits; TP<0.01mg L^{-1}). This would indicate that during larger storm events, where significant runoff is produced from the roadways, there might actually be a dilution effect occurring, which may, at least partially, explain stream P concentrations much lower than P concentrations observed from the landscapes (discussed in stream section).

Precipitation

Precipitation can be another source of P to the watershed, both to the land uses and directly to the water bodies. However, of the 68 precipitation samples collected, only 42 had DP levels above the minimum detection limit of 0.002 mg L^{-1}. Since the majority of P in precipitation is expected to be soluble, and the DP levels were low, samples were not subjected to TP analysis. Of the 42 samples with detectible DP, the mean concentration was 0.01 mg L^{-1}. Using the DP concentration and the precipitation depths measured from these 42 events, DP deposition from precipitation ranged from 0.003 to 0.020 kg P ha^{-1}.

Tree Canopy Throughfall

Forest canopies have been shown to be effective at capturing airborne particulate matter on the leaf surface and trees also produce pollen, which is high in P (*27*) and thus may be both a source and sink of P. It is not inconceivable to think that there may be P associated with the airborne particles deposited on tree leaves. Additionally, tree leaves can have P contents of 2 to 7 g kg^{-1} dry weight, which has the potential to leach out under certain conditions (*28, 29*). Throughfall precipitation from the FR canopy was measured 54 times. The vast majority of the P was in the DP form (89%), indicating that airborne particles are not contributors, or are small enough to act as a dissolved component. The mean DP concentration during the growing season was 0.16 mg L^{-1}. However, a washoff effect was apparent, as storms with precipitation greater than 25 mm had dramatically lower DP concentrations in throughfall (0.05 mg L^{-1}). In the late

Fall, Winter, and early Spring, following leaf senescence (non-growing season), the throughfall concentration was not significantly different from the concentration in precipitation (0.01 mg L^{-1}). This indicates that the leaves are indeed a source (or temporally unstable trap) for P in the ecosystem. The mass of P delivered to the soil surface as a result of throughfall on the FR land uses was estimated by multiplying the precipitation volume by the corresponding concentration. Phosphorus delivery varied based on season, from 0.004 kg ha^{-1} in the non-growing season (similar to the precipitation), to as high as 0.244 kg ha^{-1} $event^{-1}$ in the Summer (0.103 kg ha^{-1} $event^{-1}$ mean growing season DP delivery in throughfall). In some events, the runoff collected from the FR land use had lower P concentration than the throughfall collected from the canopy. This would indicate that the FR land use was removing P, likely through soil sorption, as the initially concentrated DP in the throughfall infiltrated the soil.

Following senescence and subsequent leaf drop, the P that was a component of the tree leaf tissue is subjected to breakdown and release as the tissue decays. This P is more easily transported since it is at the soil surface, and in a more mobile form. Additionally, it was observed that prior to significant breakdown of the leaf litter (December or January), the leaf mass actually appeared to reduce infiltration of precipitation into to the soil, and to increase runoff losses. The increased runoff losses, coupled with a slightly higher P concentration, resulted in increased P mass loss. However, variation masked any statistical significance.

Stream Phosphorus Load

The nutrient load in the stream was highly dependent on the sampling location, season, and discharge rate of the stream (Table IV, and Figure 4). Streamflow discharge characteristics are discussed in (*10*).

As would be expected, discharge has a highly significant relationship with the P losses from both areas of the watershed (Figure 4). High stream discharge resulted in greater P transport in the stream, particularly for PP which requires more turbulent flows to keep sediment entrained. Dissolved P loads increased as the stream flowed through the urban area, indicating that the land uses in the urban area were contributing P to the stream in excess of what would be expected from the undeveloped area of the watershed. The highest daily load was measured in August 2004, at the watershed outlet and was 0.009 kg ha^{-1}, nearly three times higher than the corresponding load at the inlet. Normalized DP loads at the watershed outlet under storm conditions were, on average, twice as high as the normalized loads at the inflow. Dissolved P loads between the undeveloped and urban areas did not differ significantly under low flow conditions. This indicates runoff is the dominant P loading mechanism in the stream. Therefore, reducing runoff P losses from the land uses may result in a reduction in the stream P load.

Table IV. Significance Table for Multiple Analysis of Variance by Multiple Regression for Stream P Loads on ln Transformed Variables

Source[a]	DP	PP	TP
		p-value	
Sampling Location	0.009	0.021	0.007
Discharge	<0.001	<0.001	<0.001
Season	0.004	0.290	0.109
Location x Discharge	0.002	0.007	<0.001

[a] Source of variation is included in the model if significant at $\alpha \leq 0.05$. Main effects are included in the model if the interaction is significant at $\alpha \leq 0.05$.

During storm events, normalized PP loads at the watershed outlet were on average two and a half times higher than the inflow, and reached as high as several orders of magnitude higher in August/September 2004 and March/April 2005 (Figure 4). During the winter months the differences were lower, possibly due to snow cover and frozen ground preventing erosive processes from removing particulate matter. Indeed, soil temperatures were well below 0°C for much of the winter (December-March) in both 2004 and 2005. The low PP losses from the upper forested area of the watershed are likely due to the low slope, low runoff, and a good forested groundcover preventing erosion. Interestingly, normalized DP, PP, and TP losses at the inflow were higher in 2003 than the following years (Figure 4). In 2003 there was wetland construction in the upper area of the watershed which resulted in considerable disturbance to the area, and was likely the cause of elevated P loads from the upper watershed. Irrespective of the stream flow volume, PP losses in the stream tended to increase with prolonged precipitation events, as seen by others (*30*). There are several explanations for this. First, saturated soil can contribute more particulate matter because of the reduced cohesive nature of saturated soil (*31*). Second, while it is not always the case, prolonged precipitation events generally increase the stream velocity, and hence the sediment and P transport capacity of the water.

Some of the increase in stream PP loads in the urban area can be attributed to land use PP loss, but some is also likely due to stream bank erosion. There are several areas in the urban area of the watershed where the stream was channelized, which reduced flooding, but increased stream velocity, and hence erosion downstream of the channelized area. Indeed, the stream bank in areas downstream from the channelized section exhibited significant erosion. Particulate P loads were about 33% of the TP load at the watershed outlet and only 10% at the inflow, indicating little erosion occurring in the upper forested area of the watershed.

Total P mass losses differed between gauging stations and were principally influenced by stream discharge (Table IV, Figure 4). The normalized TP mass

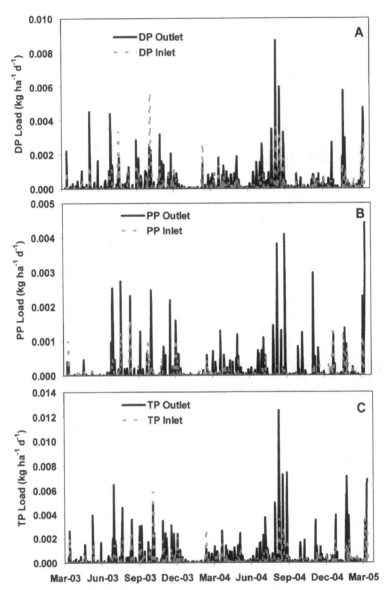

*Figure 4. Normalized dissolved phosphorus (DP) (A), particulate phosphorus
(PP) (B), and total phosphorus (TP) (C) mass losses in streamflow at the inlet
to the urban area and watershed outlet. Losses are normalized on a
kg ha⁻¹ d⁻¹ basis.*

losses at the inflow to the urban area and the watershed outlet differ rather dramatically. Total P mass losses were relatively low in the Winter, reflecting frozen conditions reducing runoff/soil interactions, as well as reduced mineralization and microbial activity. Total P losses were highest in the Summer, particularly 2004, (August/September) when the remnants of three hurricanes delivered more than 200 mm of precipitation to the region in a one-month period. During that one-month period (August/September), TP losses from the watershed were approximately 20% of the total measured P loss over the two-year study. Stream flows were nearly as high in March/April 2005 (*10*), but TP losses were half as high, likely due to the differences in mechanisms of the events (Figure 4). Prior to the August/September 2004 events, the soil was already saturated (antecedent precipitation from 1 July to 10 August 2004 was 188 mm), resulting in prolonged durations of overland flow increasing transport of sediment bound P to the stream. Saturated conditions also fostered the release of P from the soil (*18*), which in conjunction with fertilization during the Summer increased the DP available for runoff. The March/April event consisted primarily of snowmelt and rain on frozen, snow-covered ground. Nearly 130 mm of precipitation fell in a two week period (26 mm of which initially fell as snow), and runoff production was of the same order of magnitude as the August/September events (*10*). This may lead one to expect high P losses, but the soil was frozen (average temperature at 100 mm depth was -3.1°C) with snow cover for the majority of the event, which reduced the interaction of the precipitation and overland flow with the soil and reduced the erosive loss. Total P losses in the March/April event were approximately half those measured in the August/September event (Figure 4). Similarly, DP losses were also significantly lower in March/April than in August/September. However, PP losses in the March/April event were of the same order of magnitude as the August/September event (Figure 4). Particulate P losses from the land use were low in March/April, which may implicate roadways as a significant source of the P load seen in the stream during the event. There was considerable build up of road sand and other debris observed during the Winter, which was flushed from the roadways during the event, and may have increased the stream P load.

While the P load in the stream was at times high, it was not likely to be problematic for the stream itself, as turbulent waters are not prone to eutrophication because pollutants are flushed rapidly from the system. However, the stream P load is important to consider when assessing the threat of eutrophication in receiving waters (Cayuga Lake in this study). Dissolved P is probably the greatest threat because it is readily available for use by organisms such as cynaobacteria, which enhance eutrophication (*32*). However, PP can be problematic if delivered to pristine receiving waters where it can desorb from the particulate form and dissolve in the surrounding water, enhancing eutrophication (*33*). Numerous factors are involved in this process including particle settling velocity, receiving water chemistry, and stage of eutrophication. These factors are beyond the scope of this study but warrant further work.

Conclusions

A multivariate analysis of land use runoff indicated that while land use is important with respect to P loss, in most cases, it is the characteristics of the site that are most important in assessing the impact on water quality, particularly during wet years. The interaction between land use and watershed characteristics was highly significant for all variables analyzed. All three land uses monitored in this study (FL, UB, and FR) can detrimentally impact water quality under certain conditions.

From a watershed management perspective, application of P in excess of the plant requirements can, in some cases, result in higher DP loss in runoff, particularly if there is a history of excessive P applications. Therefore it may be intelligent to reconsider P application in those areas of the watershed that produce large runoff volumes or have high soil P levels. However, fertilization in this study has proven to reduce runoff volumes over the unfertilized land uses, which reduced the total mass loss of a nutrient. During the wet years of this study, DP losses from the FL land use were significantly higher than the UB and FR lands uses on the shallow low storage soils, but were equal to, or lower than, the DP losses from the UB and FR land uses on the deeper, high storage soils. Indeed, previous work in New York State (*11, 26*) has shown that fertilization at slope breaks or saturated soils can result in nutrient contamination of surface water. However, much research has also shown that fertilization can increase plant biomass and density, ultimately reducing P loss (26). Unmanaged or low maintenance land uses (i.e. abandoned areas, minimally managed areas) are a potential source of nutrients and particularly sediment. The PP losses were highest from the FR and UB land uses on the shallow soils, generally due to little or no ground cover to prevent erosive losses. Total P losses were not significantly different on the shallow soils, but were lower from the FL land use on the deeper less runoff prone soils. While there were clear land use effects on P loss in this watershed, the dominant P source areas, irrespective of land use, were the shallow low storage soils.

As the stream flowed through the urban area, there was an increase in the DP, PP, and TP load compared to the predominately undeveloped forested area of the watershed, which is attributable to the urban development. Of particular concern was the dramatic increase in PP as the stream flowed through the urban area, indicating that erosive processes were removing a considerable amount of P in the urban area.

The level of contaminant input to freshwater bodies is a function of numerous factors, many of which are interrelated and difficult to measure with certainty. However, reductions in contaminant input can be achieved through intelligent development, in conjunction with correct land management practices. Best management practices in this and many other urban watersheds should focus on reducing runoff (via stormwater management) and P loss from these areas to realize the largest reduction in P loading to surface water. Best management practices to reduce P loss from fertilized areas could include reducing or eliminating P fertilization of the turfgrass unless a soil test indicates a need, moving the typical late Fall P application to an earlier date, and avoiding P application prior to or during wet periods. Reducing P loss from the unfertilized areas should focus on reducing erosion on the land uses with little ground cover. Increasing the ground cover would be the most effective way of reducing sediment-based P losses. However, this may require fertilization, which can reduce PP loss but increase DP loss. This and many other urban watersheds could benefit from stormwater management structures, such as detention ponds or infiltration basins, to reduce storm flows and allow P laden runoff to infiltrate the soil and reduce the P load in the stream. In general, the urban area of the watershed, due to a greater abundance of impervious surfaces, shallow low storage soils, and anthropogenic factors was the dominant P loading source area in this watershed, and should be the focus of any future management strategies to improve water quality.

References

1. Owens, E.M.; S.W. Effler; S.M. Doerr; R.K. Gelda; E.M. Schneiderman; D.G. Lounsbury; C.L. Stepczuk. *J. Lake Reservoir Manag.*, **1998**, *14*, 322-333.
2. Interlandi, S.J.; C.S. Crockett. *Water Res.*, **2003**, *37*, 1737-1748.
3. Munroe, D.K.; C. Croissant; A.M. York. *Applied Geography*, **2005**, *25*, 121-141.
4. Pariente, S. *Environ. Monitoring and Assessment*, **2002**,*73*, 237-251.
5. Brezonik, P.L.; T.H. Stadelmann. *Water Res.*, **2002**, *36*, 1743-1757.
6. USGS, *National Water-Quality Assessment Program*, 2001,USGS Fact Sheet 047-01. 3 pp.
7. Hamilton, P.A.; T.L. Miller; D.N. Myers. *U.S. Geol. Surv.*, 2004, Circular 1265, pp. 1-19.
8. Honer, R. R.; J.J. Skupien; E.H. Livingston; E.H. Shaver. *Terrene Institute at U.S. Environ. Protect. Agency*, Washington D.C. 1994.
9. Waschbusch, R.J.; W.R. Selbig; R.T. Bannerman. USGS, 1999.
10. Easton, Z.M.: Petrovic, A.M. In *The Fate of Nutrients and Pesticides in the Urban Environment, Vol. 997*; Nett, M; Carroll, M.J.; Horgan, B.P.; Petrovic, A.M., Eds.; American Chemical Society, Washington, D.C., 2008.

62

11. Easton, Z.M.; A.M. Petrovic; D.J. Lisk; I.M. Larsson-Kovach. *Int. Turfgrass Sci. Res. J.*, **2005**, *10*, 121-129.
12. McIntosh, J.J. *Agron. J.*, **1969**, *61*, 259-265.
13. Sharpley, A.N. *J. Environ. Qual.* **1993**, *22*, 678-680.
14.. Murphy, J.; J.R. Riley. *Anal. Chem.*, **1962**, *27*, 37-46.
15. USEPA. *Methods for Chemical Analysis of Water and Wastes*, 1980, Method 365.1.
16. SAS Inst. Inc. *SAS/STAT User's Guide. Release V 9.1*. SAS Inst. Inc., Cary, NC, 2003.
17. Petrovic, A.M.; D.J. Soldat; J. Gruttadaurio; J. Barlow. *Internat. Turf. Sci. Res. J.*, **2005**, *10*, 989-997.
18. Linde, D.T.; T.L. Watschke. *J. Environ. Qual.*, **1997**, *87*, 176-182.
19. O'Reagain, P.J.; J. Brodie; G. Fraser; J.J. Bushell; C.H. Holloway; J.W. Faithful; D. Haynes. *Marine Pollut. Bull.*, **2005**, *51*, 37-50.
20. Jensen, M.B.; H.C.B. Hansen; N.E. Nielsen; J. Magid. *Acta. Agric. Scand. Sect.*, **1998**, *B.48*, 11-17.
21. Gaudreau, J. E.; D. M. Vietor; R. H. White; T. L. Provin; C. L. Munster. *J. Environ. Qual.*, **2002**, *31*, 1316-1322.
22. Easton, Z. M.; P. Gerard-Marchant; M. T. Walter; A. M. Petrovic; T. S. Steenhuis. *Water Resour. Res.*, **2007**, 43, paper W03413,
23. Vaithiyanathan, P.; D.L. Correll. *J. Environ. Qual.*, **1992**, *21*, 280-288.
24. Garn, H. S. *U.S. Geol. Surv. Water Res. Invest.*, **2002**, Rep. 02-4130. p.1-6.
25. Kussow, W.R. *Turfgrass Producers Int. TurfNews.*, **2003**, *27* (2), 48.
26. Easton, Z.M.; A.M. Petrovic. *J. Environ. Qual.*, **2004**, *33*, 645-655.
27. Banks, H.H.; J.E. Nighswander. *Symposium on Sustainable Management of Hemlock Ecosystems in Eastern North America*, **1999**, GTR-NE-267, 168-174.
28. Tukey, H.B. *Ann. Rev. Plant Phys.*, **1970**, *21*, 305-324.
29. Sharpley A.N. *J. Environ. Qual.*, **1981**, *10*, 160-165.
30. McDowell, R.; A. Sharpley; G. Folmar. *J. Environ. Qual.*, **2001**, *30*, 1587-1595.
31. Wondzell, S.M.; J.G. King. *Forest Ecol. Manag.*, **2003**, 178, 75-87.
32. Frossard, E.; L.M. Condron, O. A.; S. Sinaj; J.C. Fardeau. *J. Eviron. Qual.*, **2000**, *29*, 15-23.
33. Sharpley, A.N.; S.J. Smith; O.R. Jones; W.A. Berg; G.A. Coleman. *J. Environ. Qual.*, **1992**, *21*, 30-35.

Chapter 4

Nitrogen Fate in a Mature Kentucky Bluegrass Turf

Kevin W. Frank

Department of Crop and Soil Sciences, Michigan State University,
East Lansing, MI 48824

Research on nitrate-nitrogen (NO_3-N) leaching in immature
turfgrass systems indicates that, in most cases, leaching poses
little risk to the environment. This may not be true for mature
stands of turfgrass. In this study, the fate of nitrogen (N) was
examined over two years for a 10-year old Kentucky bluegrass
(*Poa pratensis* L.) turf in monolith lysimeters. Two nitrogen
rates were analyzed: 245 kg N ha^{-1} (49 kg N ha^{-1} application^{-1})
and 98 kg N ha^{-1} (24.5 kg N ha^{-1} application^{-1}). NO_3-N
concentrations in leachate for the low N rate were typically
below 5 mg L^{-1}. For the high N rate, NO_3-N concentrations in
leachate were often greater than 20 mg L^{-1}, now < than 10 mg
L^{-1}. Results emphasize the importance of long-term research to
investigate nutrient fate in turfgrass and lend support support
to the theory that older turf sites should be fertilized at a
reduced N rate to minimize the potential for NO_3-N leaching.

Research on nitrate-nitrogen (NO_3-N) leaching in turfgrass indicates that in most cases, leaching poses little risk to the environment. Extensive reviews on the environmental fate of nitrogen (N) applications have been conducted by Petrovic (1) and Walker and Branham (2). Numerous factors influence N leaching from turfgrass including: N rate and carrier (3, 4, 5, 6, 7), irrigation rate (7), rooting characteristics (4, 8, 9) and N uptake by the turf (10).

The primary concern of NO_3-N leaching from turfgrass is groundwater contamination. The United States Environmental Protection Agency (USEPA) has set a safe drinking water standard for NO_3-N of 10 mg L^{-1}. Drinking water in excess of the nitrate standard may cause detrimental health effects including blue-baby syndrome (methemoglobinemia) (11). Environmental and drinking water quality concerns have prompted several studies examining the fate of nitrogen applied to turfgrass.

Research by Morton et al (7) investigated the effects of three N fertilization rates (0, 97, and 244 kg N ha^{-1}) and two irrigation practices (as scheduled by a tensiometer to prevent drainage from the root zone, and an over-watering treatment) on a Kentucky bluegrass and red fescue (*Festuca rubra* L.) mixture established four years prior. When irrigation was scheduled using the tensiometer, the highest NO_3-N concentration in leachate was 1.24 mg L^{-1} for the 244 kg N ha^{-1} rate. For the overwatering treatment, the highest NO_3-N concentration in leachate was 4.02 mg L^{-1} for the 244 kg N ha^{-1} rate. The amount of NO_3-N leached was well below the USEPA safe drinking water standard for NO_3-N.

Gold and Groffman (5) compared the effects of different land uses on NO_3-N leaching. Leaching was compared between a home lawn turf (established six years prior), corn (*Zea mays* L.) grown for silage, a mature mixed oak-pine forest (80-120 years old), and a septic system. Nitrogen was applied to the home lawn turf and corn at an annual rate of 344 kg N ha^{-1}, and 202 kg N ha^{-1}, respectively. Nitrogen entered the septic system through household wastewater, and the forest system through natural deposition. The concentrations of NO_3-N leached were highest for the septic system and lowest for the mature forest. Leachate from the home lawn turf had NO_3-N concentrations ranging from 0.2-5.0 mg L^{-1}.

Guillard and Kopp (6) investigated leaching of ammonium nitrate, polymer-coated sulfur-coated urea and an organic fertilizer in a Kentucky bluegrass turf. Leaching losses were highest for ammonium nitrate, with an average of 17% of applied N recovered in the leachate. Leaching losses occurred primarily during the late Fall through early Spring.

Starr and DeRoo (12) applied ^{15}N labeled ammonium nitrate at a rate of 180 kg N ha^{-1}, divided into two applications, to a mixture of Kentucky bluegrass and red fescue. One year after the labeled fertilizer nitrogen (LFN) application, 64% and 73% of LFN were recovered within the system when clippings were either removed or returned, respectively. Leaching losses were low, with an average

NO_3-N concentration in leachate of 1.9 and 2.0 mg L^{-1} when clippings were either removed or returned, respectively.

Miltner et al (10) conducted a mass balance N study on Kentucky bluegrass turf. Urea nitrogen was applied at an annual rate of 196 kg N ha^{-1} divided into five applications over 38 day intervals, defined by either a Spring or Fall application schedule. NO_3-N concentrations in leachate were generally below 1 mg L^{-1} throughout the study. Only 0.23% of LFN was collected in leachate over the 2 years of the study. Total recoveries of LFN were 64% and 81% for the Spring and Fall application schedules, respectively.

The majority of N fate research has been conducted on relatively immature turf stands. The age of a turf stand has been proposed as an important factor influencing N fate. Bouldin and Lathwell (13) suggested that the ability of a soil to store organic N under relatively constant management and climatic conditions, which are typical of turf systems, would decrease with time and eventually an equilibrium level of soil organic N would be obtained. Porter et al (14) examined total N content in soil to a depth of 40 cm in 105 turf systems ranging in age from 1 to 125 years old. The data suggest that soil organic matter accumulation is rapid in the first ten years after establishment, and slowly builds to an equilibrium at 25 years, when no further net N immobilization occurs. Porter et al (14) concluded that there is a rather limited capacity of the soil to store organic N, and that after ten years the potential for over-fertilization is greatly increased.

Older turf sites, or sites with high organic matter contents, should be fertilized at a reduced N rate to minimize the potential for NO_3-N leaching (1). Petrovic (1) theorized that the rate of N applied to younger turf stands (less than ten years of age) should equal the rate at which N is used by the plants, lost to the atmosphere, and stored in the soil. Older turf sites (greater than 25 years of age) lose the ability to store additional N in the soil, and therefore should be fertilized at a rate equal to the rate N is used by the turf and lost to the atmosphere (1).

Duff et al (3) examined leaching losses of N applied to a mature Kentucky bluegrass stand (25 years). Over a 19-month period, NO_3-N concentrations in leachate were below 10 mg L^{-1} for all N fertilization rates, except for two sampling dates. After eight years of intensive management, NO_3-N concentrations in leachate were not appreciably greater than those reported for younger sites.

The current research was undertaken due to the lack of long term data on nitrogen fate in mature turfgrass stands. The research objectives were to quantify NO_3-N concentrations in leachate, and to determine the fate of LFN in clippings, verdure, thatch, soil, roots and leachate for a Kentucky bluegrass turf ten years after establishment. This research was conducted to answer the question of whether there is a greater N leaching from older mature turfgrass site if N inputs (fertilization) are not reduced with age.

Materials and Methods

Between 1989 and 1991, four monolith lysimeters were constructed according to the specifications of Miltner et al (*10*) at the Hancock Turfgrass Research Center, Michigan State University. The lysimeters, 1.14 m in diameter and 1.2 m deep, were constructed with grade 304 stainless steel (0.05 cm thick). The bottom of each lysimeter was constructed with a 3% slope to facilitate collection into a 19 L jug. In September 1990, the lysimeters and surrounding area were treated with glyphosate and then sodded with a polystand of Kentucky bluegrass (cv. 'Adelphi', 'Nassau', and 'Nugget'). Prior to the glyphosate application, the area had been a turfgrass stand for six years. Between 1991 and 1993, the lysimeters were used for a mass balance N study. From 1994 through 1997, no data were collected from the lysimeters but the turfgrass was fertilized with urea at an annual rate of 147 kg N ha^{-1}. In 1998, fertilizer treatments were initiated with two annual N rates of 98 and 245 kg N ha^{-1}. These nitrogen rates are typical of low and high rate fertilization programs for Kentucky bluegrass in Michigan (*15*). The soil type of the lysimeters and adjacent microplot area was a Marlette fine sandy loam (Fine-loamy, mixed mesic Glossoboric Hapludalfs), with a pH of 7.4. The particle size distributions were 659 g kg^{-1} sand, 227 g kg^{-1} silt, and 114 g kg^{-1} clay.

A detailed description of materials and methods is provided in Frank et al (*16*). Briefly, the turfgrass was mowed twice a week at 7.6 cm with the clippings returned. Irrigation replaced 80% of potential evapotranspiration, estimated by a WS-200 Rainbird Maxi® weather station (Rainbird, Glendora, CA). In the autumn of 2000, 90 microplots were installed in the area adjacent to the lysimeters in a completely randomized design, with four replications. The microplots were constructed of 20 cm diameter polyvinyl chloride (PVC) piping 45 cm in length. To preserve the soil structure within the microplots, the leading edge of the PVC piping was beveled and driven into the ground using a hydraulic press until it was flush with the soil surface.

On 17 October 2000, ^{15}N double-labeled urea (10 atom % excess) was applied in solution to the microplots and lysimeters, followed by 0.5 cm of irrigation. Two of the lysimeters and half of the microplots were treated at a low N rate of 24.5 kg N ha^{-1}, and the remaining lysimeters and microplots were treated at a high N rate of 49 kg N ha^{-1}. In 2001 and 2002, the lysimeters and microplots received unlabeled N in the form of urea in solution, followed by 0.5 cm of irrigation. The low N treatment was 98 kg N ha^{-1}yr^{-1}, divided into four applications of 24.5 kg N ha^{-1}. The high N treatment was 245 kg N ha^{-1}yr^{-1}, divided into five applications of 49 kg N ha^{-1}. Nitrogen application dates for both treatments were 7 May, 4 June, 3 July, and 8 October, 2001, and 8 May, 6 June, 3 July, and 15 October, 2002. The high N treatment received additional applications on 13 September in both 2001 and 2002.

Clipping samples were collected weekly from each microplot throughout the growing season for analysis. Eight microplots, four from each N treatment, were excavated by carefully digging around the perimeter of the PVC core to ensure the core was not disturbed. Microplots were collected intact on seven sampling dates: 1 November, 2000 (15 Days After ^{15}N Treatment); 1 December 2000 (45 DAT); 19 April 2001 (184 DAT); 18 July 2001 (274 DAT); 9 October 2001 (357 DAT); 20 April 2002 (549 DAT); and 17 July 2002 (637 DAT). The core was partitioned into verdure, thatch, and soil samples, all of which were dried in a convection oven for 72 hours at 60 °C.

Verdure samples included the crown and leaf portions of the plants. Thatch samples consisted of all plant material above the soil surface after verdure was removed. Soil within the thatch samples was removed by hand massaging, and then ground to a fine powder using a mortar and pestle. Thatch soil was analyzed as a soil depth. Soil was partitioned into depths of 0-5, 5-10, 10-20, and 20-40 cm. Root samples were collected and dried at 60 °C in a convection oven for 72 hours, weighed, and then ground to pass a 0.5 mm screen using a UdyMill Cyclone® Sample Mill. The ground samples of clippings, verdure, thatch, roots, and soil were dried for an additional 24 hours in a convection oven at 60 °C.

Leachate from the monolith lysimeters was collected continuously throughout the experiment. Leachate was analyzed to determine NO_3-N and NH_4-N concentrations and ^{15}N enrichment by the N diffusion technique of Moran et al (17). Due to the low ^{15}N concentration in the leachate, analysis was performed for NO_3-N and NH_4-N species combined. Total N concentration and ^{15}N enrichment in clipping, verdure, thatch, root, soil, and leachate samples were determined using a mass spectrometer. Mass of N and percent of ^{15}N recovered calculations were from Kessavalou (18).

The experiment was a completely randomized design with two replications for the lysimeters and four replications for the microplots. NH_4-N and NO_3-N concentration, LFN recovered and the percent of applied LFN recovered (%LFN) were determined for each leachate sampling date. Leachate data were analyzed as a two-factor experiment, with N rate and sampling date as factors. Soil and root LFN and %LFN data were analyzed as a three-factor experiment with N rate, DAT and depth as factors. Potential correlations between the measurements taken on the same core at different depths were accounted for by analyzing the measurements taken at different depths as repeated measures. All depths for the soil and root samples were then totaled to determine the cumulative amount of LFN and the %LFN at each sampling date. Weekly clipping data were summed to determine the cumulative amount of LFN and %LFN from all weekly sampling dates prior to the corresponding microplot sampling date. Leachate data were summed to determine the cumulative amount of LFN and %LFN from all sampling dates prior to the corresponding microplot sampling date. Kentucky bluegrass clipping, verdure, thatch, soil, root and leachate components were combined to determine the total amount and %LFN at each sampling date. The clipping, verdure, thatch, soil, root, leachate, and total recovery data were analyzed as a two-factor experiment with N rate and DAT as factors. Treatment differences were analyzed using the Proc Mixed procedure of SAS (19). When appropriate, means were separated using Fischer's LSD procedure at the 0.05 probability level.

Results and Discussion

Analysis of variance for LFN indicated that the main effect of N rate and DAT, and the N rate x DAT interaction were significant for all turfgrass components. Data are being presented by turfgrass system component: top-growth (clippings and verdure), thatch, roots, soil, leachate and total turfgrass system.

Top-Growth

The Kentucky bluegrass treated at the high N rate had a higher amount of LFN in clippings than the low N rate on all sampling dates. The total percentages of applied LFN recovered in clippings were 7 and 10% for the low and high N rate treatments, respectively (Table I).

The amount of LFN recovered in verdure was significantly different between N rate treatments at 15, 45, and 184 DAT. On these sampling dates, the high N rate treatment had a higher amount of LFN recovered than the low N rate treatment. The highest amounts of LFN recovered in verdure were from 15 to 184 DAT. The mean percentages of applied LFN recovered from 15 to 184 DAT were 15 and 18% for the low and high N rate treatments, respectively. The mean percentages of applied LFN recovered from 274 to 637 DAT were 3 and 2% for the low and high N rate treatments, respectively. The amount of LFN recovered in verdure was similar to the values reported in prior research for similar N rates applied to Kentucky bluegrass turf (10, 20). Miltner et al (10) reported the percent of applied LFN recovered in verdure declined from 23% at 199 DAT to 0.7 % at 752 DAT for a fall application of 39.2 kg N ha[-1]. These values were similar to the decline from 21% at 184 DAT to 1.0% at 637 DAT for our research.

The mean percentages of applied LFN recovered in the clippings and verdure from 15 to 184 DAT were 16 and 19% for the low and high N rate treatments, respectively.

Thatch

The amount of LFN recovered in thatch at the higher N rate was greater than at the low N rate for all sampling dates, except at 637 DAT when there was no difference (Table I). Similar to the decline in LFN recovered in verdure, the amount of LFN recovered in thatch declined after the 184 DAT sampling date. For each N rate treatment, the highest amount of LFN recovered was at 184 DAT. The mean percenaget of applied LFN recovered from 15 to 184 DAT was 12% for both the low and high N rate treatments, which corresponds to 3 and 6 kg N ha^{-1}, respectively. The mean percentage of applied LFN recovered from 274 to 637 DAT was 3% for both the low and high N rate treatments which corresponds to 0.7 and 1.5 kg N ha^{-1}.

The amount of LFN recovered in the thatch layer was less than other research (*10, 12, 20*). This could be explained by the fact that for our research, thatch soil was separated from thatch vegetative material and the LFN recovered in thatch soil was included in the overall LFN recovery of the soil.

Roots

The amount of LFN recovered for all rooting depths was summed and statistical analysis performed on the total amount of LFN recovered in roots. There were no differences in LFN recovered in roots between the low and high N rate treatments on 3 of the 5 sampling dates (Table I). On the other two sampling dates, the high N rate treatment had a higher amount of LFN recovered in roots than the low N rate treatment. The percentages of applied LFN recovered in the roots at 184 DAT were 15% (3.7 kg N ha^{-1}) and 21% (10.3 kg N ha^{-1}) for the low and high N rate treatments, respectively. The mean percentages of applied LFN recovered from 274 to 637 DAT were 9% (2.2 kg N ha^{-1}) and 11% (5.4 kg N ha^{-1}), for the low and high N rate treatments, respectively.

Soil

The amount of LFN recovered from all soil depths was summed and analyzed. There was a significant N rate x DAT interaction. The high N rate treatment had a higher amount of LFN recovered than the low N rate treatment for all sampling dates, except at 274 DAT where there was no difference (Table I). The highest amount of LFN recovered was 26.6 kg N ha^{-1} representing 54% of applied LFN at 15 DAT for the high N rate. The highest percentage of applied LFN recovered was 67%, for the low N rate treatment at 637 DAT. The mean percentages of applied LFN recovered from 15 to 184 DAT were 38 and

70

Table I. Labeled Fertilizer Nitrogen (LFN) Recovered and Percentage of Applied LFN Recovered in Kentucky Bluegrass Treated at Low and High Nitrogen Rates

N Treatment	DAT[a]	Clippings	Verdure	Thatch	Soil	Roots	Leachate	Total[b]
					kg N ha⁻¹ (%)			
Low	15	MD[c]	3.7 (15) a*[d]	2.2 (9) b*	9.7 (40) de*	MD	MD	15.5 (63) cd*
(98 kg N ha⁻¹)	45	0.3 (1) d*	3.8 (15) a*	2.1 (8) b*	7.4 (30) e*	MD	0.0 (0) a	13.5 (55) d*
	184	0.3 (1) d*	3.7 (15) a*	4.6 (19) a*	10.6 (43) cd*	3.7 (15) a*	0.1 (0.5) a*	22.8 (93) a
	274	1.2 (5) c*	1.0 (4) b	1.1 (4) c*	14.0 (57) ab	2.8 (11) b	0.1 (0.6) a*	20.2 (82) abc*
	357	1.5 (6) b*	0.9 (4) b	0.9 (4) c*	16.3 (66) a*	2.5 (10) bc*	0.1 (0.6) a*	23.4 (96) a*
	549	1.5 (6) b*	0.5 (2) b	0.8 (3) c*	12.6 (51) bc*	2.0 (8) cd	0.3 (1) a*	17.4 (71) bc
	637	1.7 (7) a*	0.3 (1) b	0.5 (2) c	16.4 (67) a*	1.5 (6) d	0.3 (1) a*	20.4 (83) ab*
High	15	MD	8.0 (16) b	5.1 (10) b	26.6 (54) a	MD	MD	39.7 (81) a
(245 kg N ha⁻¹)	45	0.8 (2) d	8.3 (17) b	6.6 (13) a	12.3 (25) e	MD	0.0 (0) e	28.0 (57) d
	184	0.8 (2) d	10.2 (21) a	6.8 (14) a	14.9 (30) de	5.1 (21) a	1.5 (3) d	39.3 (91) a
	274	3.4 (7) c	1.7 (3) c	1.9 (4) cd	15.6 (32) cd	3.0 (12) c	2.3 (5) c	27.8 (63) d
	357	4.3 (9) b	1.7 (3) c	2.4 (5) c	21.8 (44) b	3.6 (15) b	2.5 (5) c	36.3 (81) ab
	549	4.1 (8) bc	1.2 (2) c	1.5 (3) d	17.7 (36) c	1.9 (8) d	4.6 (9) b	31.1 (67) cd
	637	4.8 (10) a	0.5 (1) c	0.6 (1) e	21.1 (43) b	2.0 (8) d	5.0 (10) ab	33.9 (73) bc

[a]DAT indicates days after treatment.

[b]Total apparent ¹⁵N values are not equal to the sum of the apparent ¹⁵N recovery values for the components due to differences in statistical calculations performed on missing data points.

[c]MD indicates data was not collected.

[d]Means followed by the same letter within each column for each nitrogen rate are not significantly different at P = 0.05. Means followed by an asterisk are significantly different from the other nitrogen rate at the same date at P = 0.05. For eg., an asterisk by verdure collected at 45 DAT at the low N rate indicates that significantly less labeled fertilizer nitrogen was recovered at the low N rate than the high N rate.

36% for the low and high N rate treatments, respectively. The mean percentages of applied LFN recovered from 274 to 637 DAT were 60 and 39% for the low and high N rate treatments, respectively. The amount and percentage of applied LFN recovered in the soil were the highest among all sampling components.

Leachate

The amount of LFN recovered in leachate for the low N rate ranged from 0 to 0.08 kg N ha^{-1}, representing 0.001 to 0.32% of applied LFN (Figure 1). The

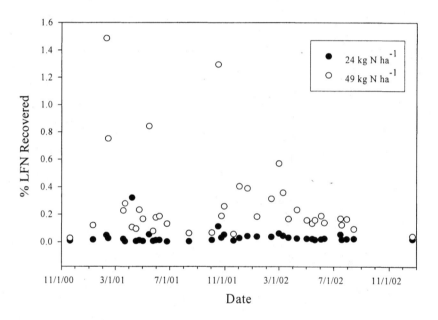

Figure 1. Percentage of labeled fertilizer nitrogen (LFN) recovered in leachate, 2000-2002.

amount of LFN recovered in leachate for the high N rate ranged from 0.01 to 0.73 kg N ha^{-1}, representing 0.03 to 1.5% of applied LFN. The total amounts of LFN recovered in leachate from 1 December 2000 through 23 December 2002 were 0.4 and 5.3 kg N ha^{-1} for the low and high N rate treatments, respectively. The total percentages of applied LFN recovered in leachate were 2 and 11% for the low and high N rates, respectively.

Previous research on the same lysimeters applied N as urea at 39 kg N ha$^{-1,}$ defined by either a Spring or Fall application schedule (*10*). The authors reported 0.2% of applied LFN recovered in leachate from a Fall application

(*10*). For our research, the percentage of applied LFN recovered was higher for both the low and high N rate treatments. The low N rate treatment had a relatively low percentage of applied LFN recovered, at 2%. However, for the high N rate treatment, 11% of applied LFN was recovered in leachate over 796 days.

Statistical analysis indicated a significant N rate x sampling date interaction for the concentration of NO_3-N recovered in leachate. From 4 June 1998 through 23 September 1999, there were no significant differences in the concentration of NO_3-N recovered in leachate between the two N rate treatments. From 1 October 1999 through 23 December 2002, the high N rate had a higher concentration of NO_3-N in leachate than the low N rate on 64 of 70 sampling dates. When there were significant differences between the N rates for the concentration of NO_3-N in leachate, the high N rate treatment always had the highest NO_3-N concentration. The NO_3-N concentration for the low N rate was less than 5 mg L^{-1} on 94 of 104 sampling dates between 4 June 1998 and 23 December 2002 (Figure 2). The flow-weighted means for the low N rate treatment ranged between a low of 2.6 mg L^{-1} and a high of 4.8 (Table II). Nitrate-nitrogen concentrations in leachate for the high N rate were greater than 20 mg L^{-1} on 20 of 104 sampling dates. The sampling dates when the NO_3-N concentration was greater than 20 mg L^{-1} were between 17 October 2001 and 12 December 2002. The flow-weighted means for the high N rate treatment ranged between a low of 5.0 mg L^{-1} to a high of 25.3 mg L^{-1}.

Our results for NO_3-N concentrations in leachate for the low N rate treatment are similar to prior research that indicates leaching of NO_3-N from turfgrass poses minimal risk to groundwater sources (*7, 10, 12*). The findings for the high N rate treatment, however, differ from previous research. The initial research on the Michigan State University (MSU) lysimeters from 1991 through 1993 found that NO_3-N concentrations in leachate from a high N rate were generally below 1 mg L^{-1} (*10*). For our research, starting in the autumn of 2001, the concentration of NO_3-N was often above 20 mg L^{-1} for the high N rate treatment. Our results support the theory proposed by Porter et al (*14*) that older turf sites should be fertilized at a reduced N rate to minimize the potential for NO_3-N leaching without negatively impacting turfgrass quality. Our results for NO_3-N concentrations in leachate for the low N rate are similar to the results of Starr and DeRoo (*12*), Morton et al (*7*), and Miltner et al (*10*), which indicate that at low N rates, leaching of NO_3-N from turfgrass poses little risk to groundwater sources. Duff et al (*3*) reported that NO_3-N concentrations in leachate were not appreciably greater for older turf sites than those reported for younger sites. The findings for the high N rate, however, differ from previous research. From 1991 through 1993, on the same site as our research, Miltner et al (*10*) reported that NO_3-N concentrations in leachate from a high N rate were generally below 1 mg L^{-1}. For the duration of our study, the concentration of NO_3-N rarely dropped below 20 mg L^{-1} for the high N rate. Our results provide

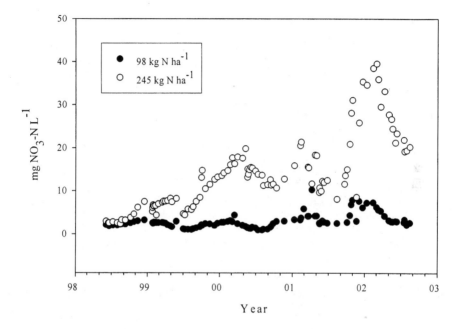

Figure 2. Nitrate-nitrogen concentration in leachate, 1998-2002.

Table II. Flow Weighted Means of Nitrate-Nitrogen Concentrations in Leachate

Year	98 kg N ha^{-1}	245 kg N ha^{-1}
1998	2.6	5.0
1999	2.0	8.5
2000	2.1	14.7
2001	3.7	18.9
2002	4.8	25.3

NOTE: units are mg L^{-1}.

support to the theory proposed by Porter et al (14) that older turf sites should be fertilized at a reduced N rate to minimize the potential for NO_3-N leaching.

Total N Recovery

Kentucky bluegrass treated at the high N rate had a higher amount and percentage of applied LFN recovered than the low N rate on all sampling dates (Table I). The highest total of LFN recovered was 39.7 kg N ha^{-1} (81%) at 15 DAT for the high N rate. Soil accounted for the highest amount of LFN among the turfgrass, soil and leachate components, regardless of N rate or sampling date. The lowest amount of LFN recovered was 13.5 kg N ha^{-1} (55%) at 45 DAT for the low N rate treatment.

As discussed previously in this chapter, there were distinct differences in the amount of LFN recovered in the turfgrass components in the first 184 days of sampling when compared to the samplings from 274 to 637 DAT. The mean total percentage of applied LFN recovered for both N rate treatments from 15 to 184 DAT was 86%. The turfgrass components (clippings, verdure, thatch and roots) contained 48% of the applied LFN. The soil contained 37% of the applied LFN and the leachate accounted for 1%. From 274 to 637 DAT, the mean total percentage of applied LFN recovered for both N rate treatments was 72%. The turfgrass components (clippings, verdure, thatch and roots) contained 20% of the applied LFN. The soil contained 50% of the applied LFN and the leachate contributed 2%. For the 15 to 184 DAT samplings, the recovery of applied LFN was high (86%) and the majority of applied LFN was recovered in the turfgrass plant. As the time from LFN application increased, the recovery of LFN decreased and the percentage of applied LFN recovered in the turfgrass plant decreased from 48 to 20%. While the percentage of applied LFN recovered in the turfgrass plant decreased from the first three samplings to the final four samplings, the percent of applied LFN recovered in the soil increased from 37 to 50%. The increase in N in the soil may be the result of N mineralization from clippings, verdure and thatch and would be available for plant uptake.

The mean total percent recovery of applied LFN for our research agreed with previous research ($10, 12, 20, 22$), but the distribution of LFN among the turfgrass and soil components differed. The amount of LFN recovered in the soil component was higher, and the amount of LFN recovered in the thatch was lower, than previous N balance research ($10, 21, 22$). The difference in turfgrass maturity between our research (10 to 13 years old) and the prior mass balance research which was either recently established or was not identified in the study might be a factor in the difference in LFN allocation. Researchers have suggested that denitrification and NH_3 volatilization losses are responsible for reported recoveries of applied LFN of less than 100%. Substantial losses of N can occur through denitrification when conditions are appropriate. Horgan et al

(*21*) compared denitrification losses from bare soil and Kentucky bluegrass. Denitrification losses (N_2 and N_2O) accounted for 7 and 19% of applied LFN for bare soil and Kentucky bluegrass systems, respectively. Mancino et al (*23*) reported little to no denitrification losses when the soil was below 80% saturation in combination with low soil temperatures; however, when saturated soil conditions were combined with high soil temperatures, denitrification losses were high. Our research returned 80% evapotranspiration twice a week, which at least temporarily may have created favorable conditions for denitrification losses. Volatilization losses may also have accounted for incomplete recovery of applied LFN, but, since the plots were irrigated immediately following N application, it is unlikely that this was a significant loss.

Conclusions

The majority of applied LFN was recovered in the soil, averaging 51% (12.5 kg N ha^{-1}) and 38% (18.6 kg N ha^{-1}) for the low and high N rates, respectively. The amount of LFN recovered in leachate from lysimeters treated at the high N rate was higher than when the turfgrass was only 1 year old (*10*). From 17 October 2000 through 23 December 2002, a period of 796 days, 2% and 11% of applied LFN were recovered in leachate for the low and high N rates, respectively. The flow-weighted means for the low N rate treatment ranged between a low of 2.6 and a high of 4.8 mg L^{-1}. The flow-weighted means for the high N rate treatment ranged between a low of 5.0 to a high of 25.3 mg L^{-1}.

The results for the low N rate were similar to the results reported by Miltner et al (*10*) at the same site from 1991-1993, and indicate that at the low N rate, the potential for groundwater contamination is minimal. At the high N rate, the amounts of LFN recovered and the concentrations of NO_3-N in leachate were substantially greater than the values reported by Miltner et al (*10*). Nitrate-nitrogen concentrations in leachate for the high N rate were greater than 20 mg L^{-1} on 20 of 104 sampling dates. The sampling dates when the NO_3-N concentrations were greater than 20 mg L^{-1} were between 17 October 2001 and 12 December 2002. This research indicates that as a turfgrass stand matures, single dose, high rate, water soluble N applications (49 kg N ha^{-1} application^{-1}) should be avoided to minimize the potential for NO_3-N leaching.

The original research on this site was conducted over a time frame of two years when the age of the turfgrass was 1 to 3 years old. The results presented in this paper are from data collected when the age of the turfgrass was 10 to 13 years old. The long term N fate research at Michigan State University is ongoing and future results will be reported. This chapter should be considered as an interim report on N fate in mature turfgrass stands.

Acknowledgements

The authors express thanks to the United States Golf Association, the Michigan Turfgrass Foundation, and the Michigan Agricultural Experiment Station for funding support. Graduate student support for Kevin O'Reilly was provided by the Paul E. Rieke graduate assistantship.

References

1. Petrovic, A.M. *J. Environ. Qual.*, **1990**, 19, 1-14.
2. Walker, W.J.; B. Branham. In *Golf Course Management and Construction-Environmental Issue;* J.C. Balogh; W.J. Walker, Ed.; Lewis Publishers: Chelsea, MI, 1992; 105-219.
3. Duff, D.T.; H. Liu; R.J. Hull; C.D. Sawyer. *Int. Turfgrass Soc. Res. J.,* **1997**, 8, 175-186.
4. Geron, C.A.; K. Danneberger; S.J. Traina; T.J. Logan; J.R. Street. *J. Environ. Qual.,* **1993**, 22, 119-125.
5. Gold, A.J.; P.M. Groffman. In *Pesticides in Urban Environments: Fate and Significance;* K.D. Racke; A.R. Leslie, Ed.; ACS symposium series 522, ACS, Washington D.C., 1993; 182-190.
6. Guillard, K.; K.L. Kopp. *J. Environ. Qual.* **2004**, 33, 1822-1827.
7. Morton, T.G.; A.J. Gold; W.M. Sullivan. *J. Environ. Qual.*, **1988**, 17, 124-130.
8. Bowman, D.C.; D.A. Devitt; M.C. Engelke; T.W. Ruffy, Jr. *Crop Sci.*, **1998**, 38, 1633-1639.
9. Jiang, Z.; J.T. Bushoven; H.J. Ford; C.D. Sawyer; J.A. Amador; R.J. Hull. *J. Environ. Qual.* **2000**, 29, 1625-1631.
10. Miltner, E.D.; B.E. Branham; E.A. Paul; P.E. Rieke. *Crop Sci.*, **1996**, 36, 1427-1433.
11. USEPA. *Technical Factsheet on Nitrate/Nitrite,* URL http://www.epa.gov/safewater/dwh/t-ioc/nitrates.html
12. Starr, J.L.; H.C. DeRoo. *Crop Sci.*, **1981**, 21, 531-536.
13. Bouldin, D.R.; D.J. Lathwell. *Bulletin 1023.* 1968, Cornell University Agriculture Experiment Station.
14. Porter, K.S.; D.R. Bouldin; S. Pacenka; R.S. Kossack; C.A. Shoemaker; A.A. Pucci, Jr.; *OWRT Project A-086-NY.* 1980, Cornell Univ., Ithaca, NY.
15. Rieke, P.E.; G.T. Lyman. *MSUE Bulletin E05TURF.* 2002, Mich. St. Univ. Coop. Ext. Serv., East Lansing, MI.
16. Frank, K.W.; K.M. O'Reilly; J.R. Crum; R.N. Calhoun. *Crop Sci.*, **2006**, 46, 209-215.
17. Moran, K.K.; R.L. Mulvaney; S.A. Khan. *Soil Sci. Soc. Am. J.*, **2002**, 66, 1008-1011.

18. Kessavalou, Anabayan. *Ph.D. thesis*, Univ. Nebraska, Lincoln, NE, 1994.
19. *The SAS system release 8.2 for Windows*. 2001, SAS Inst., Cary, N.C.
20. Frank, K.W.; R.E. Gaussoin; T.P. Riordan; W.W. Stroup; M.H. Bloom. *Int. Turfgrass Res. J.*, **2001**, 9, 277-286.
21. Horgan, B.P.; B.E. Branham; R.L. Mulvaney. *Crop Sci.*, **2002**, 42, 1595-1601.
22. Engelsjord, M.E.; B.E. Branham; B.P. Horgan. *Crop Sci.*, **2004**, 44, 1341-1347.
23. Mancino, C.F.; W.A. Torello; D.J. Wehner. *Agron. J.*, **1988**, 80, 148-153.

Chapter 5

Discharge Losses of Nitrogen and Phosphorus from a Golf Course Watershed

K. W. King[1], J. C. Balogh[2], and D. Kohlbry[3]

[1]Soil Drainage Research Unit, Agricultural Research Service,
U.S. Department of Agriculture, Columbus, OH 43210
[2]Spectrum Research, Inc., 4915 East Superior Street, Suite 100,
Duluth, MN 55804
[3]Northland Country Club, 3901 Superior Street, Duluth, MN 55804

Golf course turf accounts for approximately one million hectares of land in the United States, and is the most intensively managed system in the urban landscape. Discharge from golf course turf may potentially transport nutrients into surface water. The primary objective of this research was to assess the small watershed scale hydrologic and water quality impact from a well maintained golf course. The study site was a sub-area of Northland Country Club located in Duluth, MN. Surface water discharge and nutrient concentrations [NO_3-N, NH_4-N, dissolved reactive phosphorus (DRP), TN, and TP] were collected for a 2.5 year period (June 2002-November 2004). The mean measured rainfall/discharge coefficient during the study period was 0.46. Measured NO_3-N and NH_4-N concentrations at the inflow and outflow sites were not significantly different ($p > 0.05$), however, concentrations of TN, DRP, and TP at the same locations were significantly different ($p < 0.05$). Nutrient load attributed to the course was 0.11 kg ha^{-1} yr^{-1} NH_4-N, 0.59 kg ha^{-1} yr^{-1} NO_3-N, 0.14 kg ha^{-1} yr^{-1} DRP, 2.79 kg ha^{-1} yr^{-1} TN, and 0.27 kg ha^{-1} yr^{-1} TP. Nitrogen loads from this site pose minimal environmental concerns; however, phosphorus concentrations are consistent with concentrations known to lead to eutrophic conditions.

Introduction

Environmentally sound management of golf course turf provides both public and private facilities with environmental, cultural, and economic benefits. There are approximately 16,000 golf courses operating in the United States (1). Public demand is increasing for golf course managers to maintain high quality turf on golf courses but also to protect water and soil resources in the vicinity of these facilities (2, 3). The perception (4, 5, 6, 7, 8) and potential (9) for nutrients and pesticides to be transported in surface water is well documented. Management of existing golf courses and construction of new facilities is often a "lightning rod" of environmental and water quality concern (2). Whether or not that concern is warranted is often debated because of limited information on water quality exiting golf courses. High-quality watershed scale data are needed to adequately address this issue.

Plot scale turfgrass studies (e.g., 10, 11, 12, 13, 14) and, to some degree, watershed studies (15, 16, 17, 18, 19) have addressed runoff volume and nutrient loss from turf. The plot scale studies generally focused on small areas from plots or individual greens or fairways (20, 21), and the data is often limited to concentrations rather than loadings. Nitrate concentrations from these selected studies were generally less than 10 mg L^{-1}, while phosphorus concentrations ranged from 0.5 to 8 mg L^{-1}. Studies on small scales are valuable, but they may not represent the diversity and connectivity associated with a watershed scale turf system. The watershed scale assessments generally confirm that the concentrations reported in plot scale studies are of the same magnitude as those reported in the watershed scale studies. However, caution should be exercised when drawing linkages between the plot and watershed scale studies, because the same response variables were not always measured and reported for all the cited studies. Additionally, due to hydrologic variability, the loadings measured from watershed scale golf course facilities are generally greater than those reported from plot scale studies. Cohen et al (20) emphasize the need for more comprehensive (concentrations and loadings) field-scale water quality studies on golf courses. The primary objective of the current research was to assess the impact that golf course management has on nutrient losses. This research was designed to address the following hypothesis: that nutrient discharge losses induced by turf management from a watershed scale recreational turf system (golf course) are not significant.

Methods

Experimental Site

The selected study site for this research effort was Northland Country Club (NCC), an historic country club and high quality, private golf course located in

Duluth, MN. NCC has several subwatersheds or drainage areas with unnamed streams draining into Lake Superior. The study area is located along a stream on the northeastern part of the golf course (Figure 1). This area forms a discrete drainage area composed of 6 complete holes, three partial holes and unmanaged areas of mixed northern hardwoods and bedrock outcroppings. The 21.8 ha drainage area is comprised of 8 greens (0.3 ha), 8.5 fairways (4.0 ha), 8 tees (0.5 ha) and 17 ha of unmanaged trees and grass. The managed turf area accounts for 21.7% of the measured golf course drainage area. The drainage stream enters a natural pond located at the top of the small watershed. This stream then bisects the study area. There is a 37 meter elevation change across the study area with slopes ranging from 3 to 25%. Approximately 80 ha of low density housing and forested area feed the inflow site. A small area of typical urban housing is located on the east side of the inflow portion of this upper watershed.

NCC is located in a temperate-continental climatic region. The area is characterized by warm, moist summers and cold, dry winters. The average monthly maximum summer temperature (May-August) ranges from 16°C to 25°C (62°F to 77°F) while the average monthly maximum winter temperature (December-March) ranges from -9°C to 0°C (16°F to 32°F). Normal annual precipitation is 780 mm, half of which is generally frozen. The stream bed at the outlet is typically frozen solid from the end of November through the end of March.

Soils on the course are characteristic of lacustrine clay deposits, moderately deep (3 to 6 m) over bedrock. The Cuttre (very-fine, mixed, active, frigid Aeric Glossoaqualfs) and Ontonagon (very-fine, mixed Glossic Eutroboralfs) soils are the dominant soil series on this site with inclusions of the more poorly drained Bergland (very-fine, mixed, nonacid, frigid type Haplaquepts) series. Cuttre, Ontonagon, and Bergland soils have very similar morphological, chemical, and physical characteristics. Native vegetation associated with these soils was mixed hardwood, white spruce, and balsam fir forests. The parent material is noncalcareous clayey lacustrine deposit over calcareous clays. Perched water table conditions on the site are common, and are caused by the dense subsurface horizons and fine-textured soils.

Management on NCC (Table I) is characterized as moderate to intense. A regime of integrated management practices to control fertility, pests, irrigation and turf growth are used. These practices integrate mechanical, cultural, biological, and chemical practices designed to maintain high turf quality and limit the use of fertilizer and pesticides. NCC's integrated practices result in less dependence on chemical applications compared with many other local, private, and municipal golf courses (22).

Management practices during the study period were typical of courses in the Upper Midwest United States. Greens and tees were seeded with creeping bentgrass (*Agrostis palustris* Huds. *A. stolonifera* L.). Fairways were primarily creeping bentgrass with some Kentucky bluegrass (*Poa pratensis* L.). The roughs were a mixture of annual bluegrass (*Poa annua* L.) and Kentucky

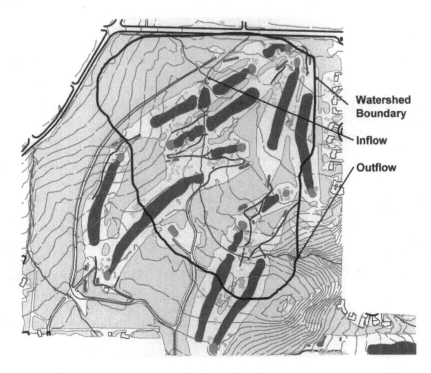

Figure 1. Layout of Northland Country Club Golf Course and study area.
(See page 1 of color inserts.)

bluegrass. NCC was irrigated with potable water from the city of Duluth. Irrigation was applied on an "as needed" basis, determined by course personnel, to replace evaporative losses. Fertilizer was applied by both dry broadcast and spray techniques throughout the year as a combination of organic, bio-stimulant, slow release, and fast release formulations. NCC uses a moderate level of nitrogen fertilizer and a small level of phosphorus fertilizer, primarily applied as slow release formulations. The number of applications in any one year is dependent on plant needs; however, the average fertilizer applications per year in 2003 and 2004 ranged from three applications on the fairways to 13 applications on the greens. Based on a review of soil test data at the golf course, Bray available phosphorus concentrations in the fairways, tees, and greens were ranked as high to very high (generally > 60 mg/kg).

Data Collection and Analysis

Surface water quantity and quality instrumentation was installed in June 2002. H-flumes (3-ft) were installed at the inflow and outflow locations of the.

**Table I. Management summary for NCC during the 2-year period
(2003-2004)**

Management Unit	Grass	Fertilizer (kg ha^{-1} yr^{-1})		Irrigation (mm yr^{-1})
		N	P	
greens	creeping bentgrass	100	60	225
tees	creeping bentgrass	240	140	240
fairways	creeping bentgrass and Kentucky bluegrass	100	60	215
roughs	annual and Kentucky bluegrass	60	40	0

study area to measure discharge. Automated Isco® samplers attached to bubbler flow meters, as well as tipping bucket rain gauges, were also added at each site to collect samples for water chemistry analysis. Automated samplers were active from April 15 to November 30, the period when the stream was generally not frozen. Temperatures can and often do exceed the freezing point during the non-sampling period; however, the durations of these 'thaw' periods were small, producing only minimal flows compared to the flows measured during the primary sampling period. Additionally, equipment limitations limited the ability to continually sample throughout the year. However, the stream was monitored on a daily basis. If flow was observed, grab samples were collected as well as stream stage. During the primary sampling period, discharge and precipitation were recorded on ten minute intervals. Discrete water samples were collected using a flow proportional approach.

Following collection, all samples were handled according to United States Environments Protection Agency (USEPA) Method 353.3 for nitrogen analysis and USEPA Method 365.1 for phosphorus analysis (23). Samples were stored below 4 °C and analyzed within 28 days. Samples were vacuum filtered through a 0.45 μm pore diameter membrane filter for analysis of dissolved nutrients. Concentrations of nitrate plus nitrite (NO_3+NO_2-N) and dissolved reactive phosphorus (PO_4-P) were determined colorimetrically by flow injection analysis using a Lachat Instruments QuikChem 8000 FIA Automated Ion Analyzer®. NO_3+NO_2-N was determined by application of the copperized-cadmium reduction, and PO_4-P was determined by the ascorbic acid reduction method (24). Total nitrogen (TN) and total phosphorus (TP) analyses were performed in combination on unfiltered samples following alkaline persulfate oxidation (25), with subsequent determination of NO_3-N and PO_4-P. From this point forward, NO_3+NO_2-N will be expressed as NO_3-N. Here, PO_4-P is used synonymously with dissolved reactive phosphorus (DRP) and will be designated from this point forward as DRP.

All statistical analyses were conducted with Minitab statistical software (26) and methods outlined by Haan (27). Normality was tested using the Kolmogorov and Smirnof test. Distributions were generally not normally distributed, thus median values were tested using the Mann-Whitney nonparametric statistic (α = 0.05).

Results and Discussion

Hydrology

Runoff occurs when the rate of precipitation is greater than the infiltration rate of the soil. The duration and total volume of discharge is directly related to precipitation and antecedent soil moisture. Increasing precipitation intensity increases the flow of runoff water and energy available for nutrient extraction and transport. The more intense the rainfall, the less time required to initiate storm runoff. Measured monthly precipitation during the study period was generally less than the long-term median amount recorded at the Duluth International Airport (Figure 2). Discharge volumes (combination of baseflow and storm event runoff) for the study period were equivalent to approximately 46% of the precipitation volume (Table II). The deep to moderately deep clayey soils on this course have some increased risk of surface runoff.

Nutrients

Periodic applications of nutrients are essential to maintaining high quality turf (28). Runoff losses of fertilizers are directly related to their timing and rate of application, formulation, chemical properties, and placement.

Table II. Measured Precipitation, Intensity, and Discharge for Upland Site, Upland Plus NCC and NCC during Data Collection Period April through November

Year (Apr-Nov)	Rainfall (P) (mm)	Max. Int. (mm/hr)	Upland Disch. (Q) (mm)	Q/P (%)	Upland + NCC Disch. (Q) (mm)	Q/P (%)	NCC Disch. (Q) (mm)	Q/P (%)
2003	353	18.5	65	0.18	83	0.24	147	0.42
2004	482	21.1	118	0.24	143	0.30	235	0.49

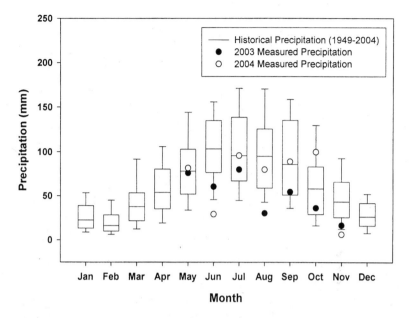

Figure 2. Historical (1949-2004) precipitation from the Duluth Airport (boxes are bound by 25th and 75th percentile values; line in the box represents the median; whiskers represent the 10th and 90th percentiles) and study period (2003-2004) measured precipitation at the experimental watershed site.

Concentrations

A range of nutrient concentrations were measured from the course (Tables III and IV). Median concentrations of NO_3-N were below 1 mg L^{-1}, and the maximum recorded concentrations were well below the USEPA drinking water standard of 10 mg L^{-1}. No statistical difference ($p > 0.05$) in median NO_3-N or NH_4-N concentration at the inflow and outflow sites was measured. This result suggests that the nitrogen fertilizer management regime used on this course does not contribute to significant increases in NO_3-N and NH_4-N in the stream. However, TN, DRP, and TP concentrations were significantly greater ($p < 0.05$) in the outflow compared to the inflow. The significant difference in TN concentration at the inflow and outflow is an indication of some organic nitrogen contribution from the course, most likely a result of tree litter decomposition and residue from the forested areas. The measured phosphorus concentrations were consistent with concentrations shown to cause eutrophic conditions in lakes, ponds, and streams (29). Increases in phosphorus concentration were generally noted with precipitation events.

Using the loads calculated for the course and the measured discharge, a concentration resulting from the course can be calculated. These resulting concentrations were 0.06 mg NH_4 L^{-1}, 0.31 mg NO_3 L^{-1}, 1.46 mg TN L^{-1}, 0.07 mg DRP L^{-1}, and 0.14 mg TP L^{-1}. These concentrations indicate that the course is adding some nutrients to the stream. The amounts of these additions are slightly greater than those resulting from the upland area.

Table III. Statistical Analysis[a] of Flow Proportional Nitrogen Concentrations (mg L^{-1}) in Surface Flow at NCC

	Surface flow concentration (mg L^{-1}) $(n = 325$ for inflow and $n=508$ for outflow)					
	NO₃		*NH₄*		*TN*	
	Inflow	*Outflow*	*Inflow*	*Outflow*	*Inflow*	*Outflow*
Mean	0.38	0.37	0.21	0.17	0.71	1.00
Median	0.25 a	0.26 a	0.02 a	0.03 a	0.62 a	1.01 b
Maximum	2.65	3.16	6.30	6.39	2.97	3.93

[a] Inflow and outflow medians by constituent were evaluated with the Mann-Whitney non-parametric test. Inflow and outflow medians for the same constituent followed by the same letter were not significantly different ($p < 0.05$).

Table IV. Statistical Analysis[a] of Flow Proportional Phosphorus Concentrations (mg L^{-1}) in Surface Flow at NCC

	Surface flow concentration (mg L^{-1}) $(n = 325$ for inflow and $n=508$ for outflow)			
	DRP		*TP*	
	Inflow	*Outflow*	*Inflow*	*Outflow*
Mean	0.05	0.09	0.08	0.10
Median	0.01 a	0.04 b	0.09 a	0.10 b
Maximum	2.42	2.59	0.23	0.55

[a] Inflow and outflow medians by constituent were evaluated with the Mann-Whitney non-parametric test. Inflow and outflow medians for the same constituent followed by the same letter were not significantly different ($p < 0.05$).

The results from this study suggest that using slow-release fertilizers and appropriate application methods mitigate the elevated potential for movement of chemicals to streams on the golf course. Additionally, maintenance of high quality turfgrass (*30*), the accumulation of thatch and organic matter in the topsoil (*31, 32*), and use of integrated best management practices also reduce the risk of nutrient losses (*33, 34, 35*).

Loads

Nutrient loadings (the mass of nutrient transported in surface flow) from NCC were calculated from the concentration data and the measured runoff from the course. Nutrient load was calculated by multiplying the analyte concentration by the measured water volume for that respective sample and summing over the study duration. The volume of water associated with any one sample was determined using the midpoint approach; the midpoint between each sample was determined and the volume of water calculated for that duration. The analyte concentration was assumed to be representative over that specific flow duration. Nutrient load attributed to the course was 0.11 kg ha^{-1} yr^{-1} NH$_4$-N, 0.59 kg ha^{-1} yr^{-1} NO$_3$-N, 0.14 kg ha^{-1} yr^{-1} DRP, 2.79 kg ha^{-1} yr^{-1} TN, and 0.27 kg ha^{-1} yr^{-1} TP. The loadings from this golf course are generally greater than or similar to loadings reported for native prairies (36) and forested catchments (15, 37), but less than loadings reported for agriculture (38, 39), the exception being phosphorus (Table V). Despite the relative immobility of phosphorus in soil (33), the results of this study suggest that this course may have the potential for small but significant contributions of phosphorus to surface water. This course has a long history of phosphorus applications. Once the soils become saturated with precipitated phosphorus, any additional phosphorus is more readily available for loss in surface runoff (42).

Summary and Conclusions

1. Surface water hydrology and nutrient concentrations were measured for 2.5 years at NCC.
2. A range of nutrient concentrations were detected in the surface water.
3. The nitrogen fertilization regime used on this course appears to pose little risk for significant inorganic nitrogen transport in surface runoff.
4. The measured phosphorus concentrations indicate the need for thorough soil sampling prior to additional phosphorus application. This includes characterization of soils saturated with precipitated phosphorus.
5. Nitrogen and phosphorus loadings from this course were generally greater than or similar to losses from native prairies and forests but less than loadings reported for agriculture.

Acknowledgements

The authors would like to express their gratification to Sara Beth Scadlock, Ivy Leland, and Emily Burgess for aiding in site development and design as well

Table V. Nutrient loads (kg ha^{-1} yr^{-1}) from NCC and Other Selected Land Uses

Reference	Land use	Area	NH$_4$	NO$_3$	TN	DRP	TP	Dur.	Study Site
(40)	Tifway bermudagrass	25.2 m^2	---	3.05	---	---	---	4-yrs	Griffin, GA
(10)	80% Kentucky bluegrass; 20% perennial ryegrass	37.2 m^2	0.35	0.90	---	0.12	---	18-mos	Ithaca, NY
(36)	Native prairie	89.6 m^2	0.14	0.12	0.84	0.02	0.11	5-yrs	Big Stone County, MN
(41)	golf green (bermudagrass)	0.025 ha	---	0.52	---	---	---	3-mos	College Station, TX
	golf frwy. (bermudagrass)	1.57 ha	---	0.96	---	---	---		
(16)	golf course: storm events	29 ha	---	2.1	---	0.3	---	13-mos	Austin, TX
	golf course: baseflow		---	4.3	---	0.05	---		
(17)	golf course	29 ha	---	4.0	---	0.66	---	5-yrs	Austin, TX
(15)	golf course	53 ha	1.7	3.7	13.5	1.6	3.04	2-yrs	Japan
	forest	23 ha	0.2	4.1	5.4	0.03	0.13		
(38)	95% agriculture; 5% urban	327 ha	0.34	20.4	---	0.28	1.13	1-yr	Fayette County, KY
	43% agriculture; 57% urban	506 ha	0.95	10.8	---	0.12	1.14		
	99% urban; 1% agriculture	226 ha	0.52	5.97	---	0.07	0.66		
(39)	agriculture	214 ha	1.15	5.68	25.6	0.07	3.72	3-yrs	Westmoreland County, VA
This Study	Golf course	21.8 ha	0.11	0.59	2.79	0.14	0.27	2-yrs	Duluth, MN

as routine data collection of both hydrology and water quality samples. The authors would also like to acknowledge and sincerely thank Ms. Ann Kemble for her technical support with respect to sample preparation and to Mr. Eric Fischer for providing his analytical expertise to this research effort. Additionally, the authors would like to thank the members and staff of Northland Country Club for granting us permission to conduct this study on their course. Finally, we would like to thank the U.S. Golf Association Green Section for funding support provided toward this research.

References

1. National Golf Foundation. *Golf Facilities in the U.S*; National Golf Foundation: Jupiter, Florida, 2003.
2. Balogh, J.C.; Leslie, A.R.; Walker, W.J.; Kenna, M.P. Development of integrated management systems for turfgrass. In *Golf course management and construction: Environmental issues*. Balogh, J.C.; Walker, W.J., Ed. Lewis Publishers, Inc.: Ann Arbor, Michigan, 1992; pp 355-439.
3. Beard, J.B.; Green, R.L. *J. Environ. Qual.* **1994**, *23*, 452-460.
4. Kohler, E.A.; Poole, V.L.; Reicher, Z.J.; Turco, R.F. *Ecological Engineering.* **2004**, *23*, 285-298.
5. Shuman, L.M. *J. Environ. Qual.* **2002**, *31*, 1710-1715.
6. Peacock, C.H.; Smart, M.M.; Warren-Hicks, W. In *Proceedings of the EPA Watershed 96 Conference.* U.S. EPA: Washington, D.C. 1996; pp 335-338.
7. Smith, A.E.; Bridges, D.C. *Crop Sci.* **1996**, *36*, 1439-1445.
8. Pratt, P.F. *Council for Agricultural Science and Technology, Report No. 103.* 1985. Ames, Iowa, pp 1-62.
9. Balogh, J.C.; Walker, W.J. *Golf Course Management and Construction: Environmental Issues*; Lewis Publishers: Ann Arbor, Michigan, 1992.
10. Easton, Z.M.; Petrovic, A.M. *Golf Course Management.* **2005**, *73*(5), 109-113.
11. Gaudreau, J.E.; Veitor, D.M.; White, R.H.; Provin, T.L.; Munster, C.L. *J. Environ. Qual.* **2002**, *31*, 1316-1322.
12. Cole, J.T.; Baird, J.H.; Basta, N.T.; Huhnke, R.L.; Storm, D.E.; Johnson, G.V.; Payton, M.E.; Smolen, M.D.; Martin, D.L.; Cole, J.C. *J. Environ. Qual.* **1997**, *26*, 1589-1598.
13. Linde, D.T.; Watschke, T.L. *J. Environ. Qual.* **1997**, *26*, 1248-1254.
14. Morton, T.G.; Gold, A.J.; Sullivan, W.M. *J. Environ. Qual.* **1988**, *17*, 124-130.
15. Kunimatsu, T.; Sudo, M.; Kawachi, T. *Water Sci. Tech.* **1999**, *39*, 99-107.
16. King, K.W.; Harmel, R.D.; Torbert, H.A.; Balogh, J.C. *J. Am. Water Resour. Assoc.* **2001**, *37*, 629-640.
17. Winter, J.G.; Dillon, P.J. *Environ. Pollut.* **2005**, *133*, 243-253.

90

18. King, K.W.; Balogh, J.C.; Hughes, K.L.; Harmel, R.D. *J. Environ. Qual.* **2007**, 1021-1030.
19. King, K.W.; Balogh, J.C.; Harmel, R.D. *Environ. Poll.* **2007**, (in press, doi:10.1016/jenvpol.2007.01.038)
20. Cohen, S.; Svrjcek, A.; Durborow, T.; Barnes, N.L. *J. Environ. Qual.* **1999**, *28*, 798-809.
21. Kenna, M.P. *USGA Green Section Record.* 1995, 33(1), 1-9.
22. Balogh, J.B. *Water Quantity and Quality Assessment for the Proposed Spirit Ridge Golf Course.* Spectrum Research, Inc., Duluth, Minnesota, 2001.
23. USEPA. *Methods for Chemical Analysis of Water and Wastes*; U.S. Environmental Protection Agency, EPA-600/4-79-020. Cincinnati, Ohio, 1983.
24. Parsons, T.R.; Maita, Y.; Lalli, C.M. *A Manual of Chemical and Biological Methods for Seawater Analysis*; Pergamon Press, Oxford, 1984.
25. Koroleff, J. Determination of Total Phosphorus by Alkaline Persulphate Oxidation. In *Methods of Seawater Analysis.* Grasshoff, K.; Ehrhardt, M.; Kremling, K., Eds. Verlag Chemie, Wienheim, 1983; pp 136-138.
26. Minitab Inc. *Minitab statistical software, Release 13. User's Guide: Data Analysis and Quality Tools*; State College, Pennsylvania, 2000.
27. Haan, C.T. *Statistical Methods in Hydrology*, 2nd Ed.; The Iowa State Press: Ames, Iowa, 2002.
28. Branham, B.E.; Kandil, F.Z.; Mueller, J. *USGA-Green Section Record.* 2005, 43:26-30.
29. Sharpley, A.N.; Rekolainen, S. Phosphorus in agriculture and its environmental implications. In: *Phosphorus Loss from Soil to Water.* Tunney, H.; Carson, O.T.; Brooks, P.C.; Johnston, A.E., Eds. CAB International: Wallingford, England, 1997; pp 1-53.
30. Watschke, T.L. *Golf Course Mgt.* **1990**, *2*, 18, 22, and 24.
31. Taylor, D.H.; Blake, G.R. *Soil Sci. Soc. Am. Proc.* **1982**, *46*, 616-619.
32. Zimmerman, T.L. *The effect of amendment, compaction, soil depth, and time on various physical properties of physically modified Hagerstown soil.* Ph.D. Dissertation, The Pennsylvania State Univ., University Park, PA, 1973.
33. Walker, W.J.; Branham, B. Environmental impacts of turfgrass fertilization. In *Golf course management and construction: Environmental issues.* Balogh, J.C.; Walker, W.J., Eds. Lewis Publishers, Inc.: Ann Arbor, Michigan, 1992; pp 105-219.
34. White, R.W.; Peacock, C.H. Peacock. *Int. Turf. Soc. Res. Journal.* **1993**, *7*, 1000-1004.
35. Engelsjord, M.E.; Branham, B.E.; Horgan, B.P. *Crop Sci.* **2004**, *44*, 1341-1347.
36. Timmons, D.R.; Holt, R.F. *J. Environ. Qual.* **1997**, *6*, 369-373.
37. Graczyk, D.J.; Hunt, R.J.; Greb, S.R.; Buchwald, C.A.; Krohelski, J.T. *Hydrology, nutrient concentrations, and nutrient yields in nearshore areas*

of four lakes in northern Wisconsin, 1999-2001. USGS Water-Resources Investigations Report 03-4144, U.S. Geological Survey, Reston, Virginia, 2003, pp 1-73.
38. Coulter, C.B.; Kolka, R.K.; Thompson, J.A. *J. Am. Wat. Res. Assoc.* **2004**, *40*, 1593-1601.
39. Inamdar, S. P.; Mostaghimi, S.; McClellan, P.W.; Brannan, K. M. *Trans. of the ASAE.* **2001**, *44*, 1191–1200
40. Schwartz, L.; Shuman, L.M. *J. Environ. Qual.* **2005**, *34*, 35-358.
41. Birdwell, B. *Nitrogen and chlorpyrifos in surface water runoff from a golf course.* M.S. Thesis, Texas A&M University, College Station, TX, 1995.
42. Cox, F.R.; Hendricks, S.E. *J. Environ. Qual.* **2000**, *29*, 1582-1586.

Chapter 6

The Effects of Soil Phosphorus and Nitrogen and Phosphorus Fertilization on Phosphorus Runoff Losses from Turfgrass

Douglas J. Soldat[1], A. Martin Petrovic[2], and Harold M. van Es[3]

[1]Department of Soil Science, University of Wisconsin at Madison, Madison, WI 53706
Departments of [2]Horticulture and [3]Crop and Soil Sciences, Cornell University, Ithaca, NY 14853

Turfgrass accounts for a large percentage of land in urban and suburban areas and thus, it is important to understand the effects of turfgrass on surface water quality. Runoff from natural and simulated rain events was collected from a cool season turfgrass mixture on an undisturbed sandy loam soil for two years. Field plots with runoff collectors had different soil P levels as a result of prior fertilization practices. The treatments for this study were fertilizer application levels and included no fertilizer, nitrogen (N) only, phosphorus (P) only, or both N and P. Runoff volumes were measured and a subsample was saved for dissolved and total P analysis. Phosphorus losses were 0.05% of fertilizer applied for both dissolved and total P. Low mass losses of dissolved P were observed (<0.05 kg ha^{-1} yr^{-1}) and can be attributed to the small amount of precipitation that became runoff. Application of N or P did not affect the amount of runoff from natural or simulated events. Fertilization with N or P increased the concentration of P in the runoff to a similar extent. Soil P levels had no effect on runoff P concentrations or mass losses, despite Morgan extractable soil P levels ranging from 3.7 to 35.6 mg kg^{-1} at the 0–5 cm depth. The simulated events supported the data observed from the natural events in most cases. Significant differences in infiltration rate among treatments were found on 2 of the 6 simulation dates.

Significant differences in P loss were only observed when no precipitation fell between a fertilization event and a simulated runoff event. The results of this study suggest that fertilization of established turfgrass does not result in a reduction in runoff volume when visual quality responses to N are similar to those observed in this study. Since increased P runoff concentrations were associated with N and P fertilization, the environmental impacts of turfgrass could be reduced by withholding or limiting N and P fertilizers under these conditions. Soil P level was not a good indicator of P concentration in runoff for this sandy loam and across the range of soil P levels seen in this study. Thus, predictions of runoff P loss based on soil P levels may be misleading or inaccurate for turfgrass areas.

Runoff from urban areas is the third leading source of water quality impairment in rivers, lakes, and streams in the United States (US) (1). One component of urban runoff includes nutrient loss from turfgrass areas such as golf courses, parks and home lawns. It has been recently estimated that 1.8% of the land area in the US is turfgrass (2). It is therefore of great importance to understand the effects of turfgrass management practices on surface water quality. However, relative to the large area of turfgrass in the US, little research has been conducted on the factors that influence runoff losses and runoff water quality from turfgrass.

Linde et al (3) found that creeping bentgrass reduced runoff losses compared with perennial ryegrass when both were mown at fairway height. The authors associated the reduction in runoff volume from creeping bentgrass with its greater shoot density, which allowed for increased water infiltration. Easton and Petrovic (4) found that fertilization of a mixture of Kentucky bluegrass (*Poa pratensis*) and perennial ryegrass (*Lolium perenne*) during establishment decreased runoff losses because shoot density was increased, leading to a reduction in runoff volume from the fertilized plots. Gross et al (5) seeded tall fescue (*Festuca arundinacea*) at different rates to achieve a range in shoot density, then used simulated rainfall to force runoff from the plots. They found no difference in runoff volume for shoot densities ranging from $8.7 - 56.9$ tillers dm^{-2}, a range on the low end of commonly observed turfgrass densities. Kussow (6) observed a $47 - 59\%$ reduction in runoff P losses from turf fertilized with N and P compared to an unfertilized control. However, because growth responses to P fertilizer are rarely observed in practice, it is likely that P losses could be reduced further by applying nitrogen only.

In response to poor or declining surface water quality in urban areas, Minnesota and Dane Co., Wisconsin have banned P fertilizer applications to turfgrass areas unless a soil test shows that the nutrient is required. However, as soil test recommendations for turfgrass sites typically are based on plant response, they do not consider potential environmental impacts. Furthermore, the loss of sediment from turfgrass has been shown to be minimal even under low maintenance conditions (5), suggesting the contact between runoff water and soil is reduced. Sharpley (7) concluded that crop canopy leaching could account for a significant portion of P in runoff in cotton. For a turfgrass situation, where runoff interaction with soil is reduced at the expense of an increased interaction with turfgrass tissue, it is unclear how much soil P levels influence runoff P losses.

A better understanding of nutrient runoff losses from turfgrass is needed to further improve fertilizer recommendations and soil test interpretations for turfgrass areas. This study had two objectives: 1) examine the effects of N and P fertilization of established turfgrass on runoff P losses; and 2) examine the effect of soil P level on P runoff losses from established turfgrass.

Materials and Methods

The study was initiated in May 2004 at the Cornell University Turfgrass and Landscape Research Center in Ithaca, NY. The experimental plots were 1.8 m long by 0.9 m wide and were situated on a hillside with a slope ranging from 4.4 to 9.6 %. A runoff collection unit was installed at the base of each plot following the method of Cole et al (8). Steel borders 2.5 mm thick and 10 cm wide were installed in the soil to a depth of 8 cm around the perimeter of each plot (except on the downside) for hydrologic isolation. In Spring 2005, the steel borders were replaced with common plastic garden edging.

The plots were situated on an Arkport sandy loam (620 g kg^{-1} sand, 260 g kg^{-1} silt, 120 g kg^{-1} clay) having a pH of 5.5 and an organic matter content of 27 g kg^{-1}. The site was not compacted nor severely modified before turfgrass establishment as commonly happens to yard areas during house construction; therefore, these results are more directly applicable to relatively undisturbed sites such as cemeteries, commercial lawns, golf courses, and parks.

The turfgrass growing on the site was a mixture of perennial ryegrass (*L. perenne*) and Kentucky bluegrass (*P. pratensis*). The turfgrass was established in 2000 as part of an earlier nutrient runoff study (4). During that study, the researchers applied various rates and sources of P fertilizer, which resulted in a wide range of soil P levels across the plots (Figure 1). Plots were mowed at a height of 6.3 cm as needed; clippings were returned; 2 cm of supplemental irrigation was applied once in August 2005 to prevent dormancy.

To study the effects of N and P fertilizer on P runoff losses, we employed a 2 x 2 factorial design with the four treatments consisting of two N (0 and 200 kg ha^{-1} yr^{-1}) and P (0 and 50 kg ha^{-1} yr^{-1}) fertilizer levels; treatments were replicated six times. Eight equal-sized fertilizer applications were made over the course of the two year study on the following dates: 16 May, 5 July, 9 September and 9

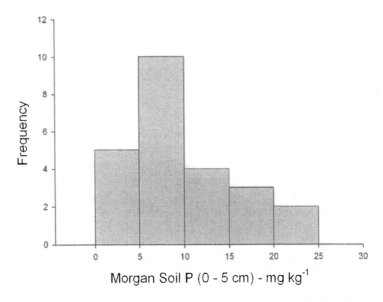

Morgan Soil P (0 - 5 cm) - mg kg^{-1}

Figure 1. Frequency distribution of Morgan extractable soil P levels of research plots. Data from samples taken on 28 September 2005.

November 2004, and on 9 June, 22 July, 23 September, and 11 November 2005. Prior to this study, the plots were last fertilized in August 2001. Nitrogen was applied in the form of sulfur-coated urea [39-0-0 (Lesco, Strongsville, OH)], and P as triple super phosphate (0-45-0). Each material was applied individually using a handheld shaker bottle.

To supplement the data from natural runoff events, runoff was generated periodically using a miniature rainfall simulator (Figure 2). Conventional rainfall simulators apply water at a constant rate which causes runoff from each plot to vary depending on the infiltration rate of the soil. The rainfall intensity of the simulator used in this study can be quickly adjusted to achieve a relatively constant runoff rate from each plot. We adjusted rainfall intensities to achieve a runoff rate of 50 mm hr^{-1} and collected the first 7.5 mm of runoff for analysis. This rate and depth were found to be typical of an average runoff event for a small, urban watershed in central New York (9).

Runoff volumes were recorded by tipping buckets outfitted with dataloggers. If the runoff depth from a rainfall event exceeded 0.1 mm, a subsample was saved and stored at 4°C for dissolved and total P analysis. Phosphorus load was calculated for individual runoff events as the product of the subsample P concentration and the runoff volume recorded by the datalogger. Total P was determined colorimetically using stannous chloride, following a persulfate digestion. For determination of dissolved P, each runoff sample was

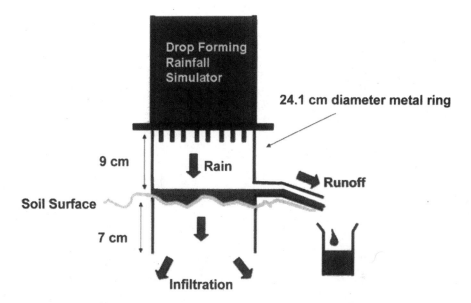

Figure 2. Diagram of rainfall simulator used in this study. More information available in Ogden et al (10). (See page 1 of color inserts.)

centrifuged for 10 minutes at 2500 rpm and P was measured in the supernatant using the ascorbic acid method (*11*).

Visual turf quality is a subjective measure of color, density, and uniformity accepted and widely practiced by turfgrass researchers in field research settings (*12*). Turfgrass quality was assessed monthly during the study in an attempt to link differences in runoff with the visual appearance of the turf in the plots. A rating scale of 1 to 9 was used, with 1 being completely brown or dead turf and 9 being the highest quality turf possible. A density count was taken near the end of the study to measure the treatment effects on shoot density over the course of the study. Three plugs were taken with a golf hole cutter (10.2 cm diameter) and all shoots were counted.

Soil samples were taken to a depth of 0-5 cm and analyzed for Morgan (*13*) extractable P on 1 May 2004, 9 December 2004, 25 May 2005 and 28 September 2005.

Statistical Analysis

Small plot runoff typically exhibits a high degree of spatial variation. For this reason, it is important to take steps to minimize this variation to accurately

98

detect treatment effects. Therefore, a spatially balanced incomplete block design with 6 replications was utilized (Figure 3). Each incomplete block consisted of 2 adjacent plots. This arrangement was found to be the most effective layout for minimizing error in mean square and average coefficient of variation (*14*). Dummy variables were assigned to each plot and spatially balanced top to bottom and left to right across the hillside to give similar comparison distances within and among the randomly assigned treatments.

Treatment effects were determined by analysis of variance using the general linear model in the SAS software package (SAS Institute Inc, Cary, NC). Type III sums of squares were used to determine significance of model parameters. Linear regression analysis was performed by the SigmaPlot graphing software (SPSS Inc, Chicago, IL).

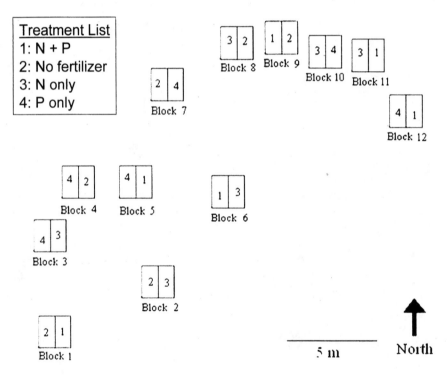

Figure 3. Layout of incomplete blocks and treatment locations on hillslope.

Results and Discussion

Natural Runoff Events

The study site had very low runoff potential. During the study, 18 runoff events from natural precipitation were observed. Precipitation averaged 1052 mm yr^{-1} over the study period, slightly greater than normal for Ithaca, NY (948 mm yr^{-1}). The amount of precipitation associated with the 18 runoff events was 533 mm. The average amount of runoff from the plots over the same time period was 3.2 mm, meaning only 0.6% of precipitation from runoff-causing storm events became runoff, or only 0.2% of total precipitation became runoff. The results of the study should be interpreted with these facts in mind. Sites with low runoff potential are not uncommon in the northeastern USA and our results are applicable to these areas.

Although 18 runoff events (an event is defined as when runoff was observed from at least 1 of the 24 runoff plots) occurred, only 3 times was runoff observed on more than 80% of the plots, thus preventing individual statistical analysis of most runoff events. Therefore, runoff depths and mass P losses were summed, and P concentrations were averaged for statistical analysis.

Total runoff losses were not affected significantly by either N or P fertilization (Table I). A possible explanation for this was that turf quality differences among the treatments were observed on only 4 of the 11 dates. On these four dates, the plots receiving N had significantly greater turf quality ratings than the others. The greatest difference was observed on 28 June 2005, when plots with N had a turf quality rating 1.2 units higher than plots that did not receive N. For the unfertilized control plots, turf quality was below 6.0 on only 2 of the 11 dates.

Table I. ANOVA Table for Total Runoff Depth, and Average Dissolved (DP) and Total P (TP) Concentrations and Mass Loss in Runoff

Source of Variation	D.F.[a]	Runoff	DP Conc.	TP Conc.	DP mass loss	TP mass loss
		-------------------------		p-value		-------------------------
Block	11	0.08	<0.01	0.29	0.04	0.31
Soil P level	1	0.58	0.58	0.85	0.26	0.71
N level	1	0.99	0.03	0.26	0.17	0.48
P level	1	0.27	0.02	0.66	0.08	0.75
N x P level	1	0.30	0.46	0.27	0.53	0.75

[a] Degrees of Freedom

On 26 September 2005, no significant differences in shoot density were found among the treatments (mean shoot density = 123 tillers m^{-2}). It can be concluded that over the 2 year study period, N fertilization did not create large differences in shoot density, and therefore fertilization effects on runoff depth were not observed.

Although differences in runoff depth were not influenced by fertilization practices (p = 0.27 and 0.99), they were influenced (p=0.08) by block (Table I). Plots near to each other tended to have similar runoff depths, although no clear trends based on slope, soil texture, or spatial location in runoff could be identified, as has been previously shown (*15*).

Dissolved P concentrations were significantly affected by plot location (block), P, and N fertilization (Table I). Interestingly, N fertilization increased the average dissolved P concentration in the runoff compared to the non-fertilized plots (Table II). Although clipping yield and tissue P content were not measured during this study, it is possible that N fertilization increased clipping yield compared to the unfertilized control, resulting in a greater amount of P in tissue to interact with runoff water. Petrovic et al (*16*) found that applying N fertilizer increases clipping yield, as well as tissue P content.

Similarly, but less surprisingly, significantly greater P runoff concentrations were measured for plots fertilized with P than those that were not (Table II). The source of P from these plots could be from direct fertilizer losses, an increase in tissue P content of the turfgrass or an increase in clipping yield. The latter two explanations are unlikely, based on the soil P levels of the test site being adequate for maximum growth (*16*). No significant difference existed between the P concentrations in runoff from turfgrass fertilized with N or P. Total P concentrations were not significant for any of the model parameters (Table I).

Dissolved and total P runoff mass losses were small (DP < 0.1 kg ha^{-1}; TP < 0.14 kg ha^{-1}) and no significant differences were detected among treatments (Table II). Mass loss is the product of runoff volume and P concentration in the runoff. In this study, treatment differences in runoff were not detected (p = 0.27 and 0.99), but treatments significantly influenced P concentration. Differences in mass losses were not detected because the large amount of error associated with runoff volume (Table II) masked the differences in concentration among treatments. Dissolved P losses from plots receiving P accounted for 0.2% of the P applied. When corrected for the P loss from unfertilized control plots, dissolved P losses were 0.04% of P applied. Total P losses from plots receiving P accounted for 0.05% of applied P when P in unfertilized control plots was accounted for. These losses are similar to other studies that reported runoff P losses from turfgrass from natural events. Kussow (*6*) observed no increase in P loss due to P fertilizer application, and an expected annual P loss of 0.2 to 1.3 kg ha^{-1}; however P fertilizer was always applied with N fertilizer, thus confounding the results. Easton and Petrovic (*4*) found runoff losses of P fertilizer ranged from less than the unfertilized control to 0.6% of P applied for various P

Table II. Effect of N and P Fertilization on Mean Dissolved P (DP) and Total P (TP) Concentrations and Mass Losses in Runoff Water (n=18)

Treatment kg ha⁻¹ yr⁻¹	Runoff mm (s.e.)	DP Conc. mg L⁻¹ (s.e.)	TP Conc. mg L⁻¹ (s.e.)	DP mass loss kg ha⁻¹ (s.e.)	TP mass loss kg ha⁻¹ (s.e.)
Nitrogen applied					
0	3.14 a	2.09 b	2.83 a	0.063 a	0.099 a
	(0.48)	(0.17)	(0.85)	(0.011)	(0.033)
200	3.22 a	2.75 a	4.37 a	0.089 a	0.136 a
	(0.43)	(0.15)	(0.76)	(0.010)	(0.29)
P₂O₅ applied					
0	2.84 a	2.10 b	3.86 a	0.061 a	0.110 a
	(0.40)	(0.14)	(0.71)	(0.010)	(0.028)
50	3.52 a	2.74 a	3.34 a	0.093 a	0.125 a
	(0.46)	(0.16)	(0.79)	(0.011)	(0.031)

NOTE: Similar letters within columns are not statistically different according to Fisher's LSD (p=0.05)

sources when compared to unfertilized control plots during the first two years after seeding. In their study, N was always applied with P which allowed for the fertilized plots to establish quicker and thus reduced runoff losses compared to the unfertilized control plots. Gross et al (17) found dissolved PO_4-P losses ranged from 0.007 to 0.12 kg ha⁻¹ yr⁻¹ from plots that did not receive any P fertilizer. The differences in mass P loss were explained by the amount of rainfall in a given year. Similarly, we found that dissolved and total P load was more closely related to runoff volume than P concentration (Figure 4).

We did not observe a reduction in P loss due to N application. Researchers have reported greater P losses in studies where rainfall simulators are used to force runoff immediately following a fertilizer application. Schuman (18) reported 10 to 11% of applied P was lost from turfgrass when runoff was forced 4 hours after P application; and Linde and Watschke (19) reported runoff P losses of 11% of P applied when runoff was forced 8 hours after the application.

Although the range in Morgan extractable soil P levels represented over 60% of the variation in soil P levels of lawns and athletic fields in New York State (20), extractable soil P level had no influence on the average concentration of P in the runoff (Table I) or the total mass loss of P (Figure 5). Similarly, Barten and Janke (21) found a poor relationship ($r^2 = 0.14$) between soil P and dissolved P in runoff from turfgrass over a very wide range of soil P levels (5-65 mg kg⁻¹ Bray-1 P). By contrast, research on agricultural fields has shown that dissolved P concentrations in runoff tend to be strongly correlated with soil P

102

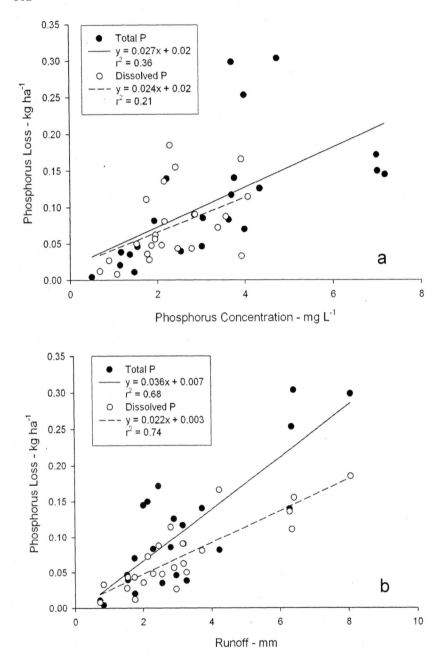

Figure 4. Relationship between (a) dissolved or total P concentration and dissolved or total P mass loss and (b) relationship between runoff depth and dissolved or total P mass loss.

levels, especially within a soil type (22). However, significant sediment loss is rarely observed from established stands of turfgrass (3, 4, 5, 17), suggesting a reduced interaction between the runoff and soil.

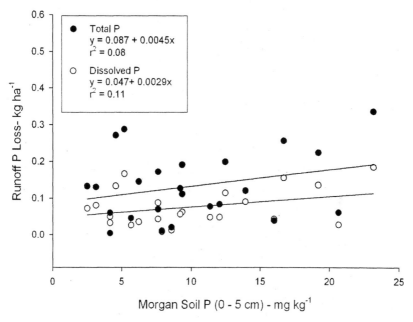

Figure 5. Relationship between soil P level and average runoff total P and dissolved P mass losses for natural runoff events.

Simulated Runoff Events

Over the course of the study, 6 runoff events were simulated to supplement the data from the natural runoff events. We were able to generate runoff from each plot on all simulation dates, allowing for the data from each date to be analyzed individually. During rainfall simulation, data were collected to allow us to calculate the infiltration rate of each plot, which served the same purpose as runoff depth from the natural runoff events.

Statistically significant differences in infiltration rate among the treatments were detected on 2 of the 6 simulation dates (Table III). In both cases, the N+P treatment had a significantly greater infiltration rate than the other three treatments which were not significantly different from each other. Also, on each of these dates the plot location was statistically significant (Table III). These findings agree with the results of the natural runoff events where plot location was found to be the most significant factor in determining runoff depth.

An equal amount of runoff (7.5 mm) was collected from each plot in order to compare treatment differences in P concentration. With one exception, no

Table III. Analysis of Variance Table for Infiltration Rate for Simulated Events

Source of Variation	D.F.[a]	28 July 2004	23 Sept. 2004	27 June 2005	23 July 2005	24 Sept. 2005	27 Sept. 2005
		---------------------------p-value---------------------------					
Block	11	0.460	0.194	0.023	0.035	0.758	0.496
N Level	1	0.464	0.405	0.121	0.038	0.351	0.145
P Level	1	0.788	0.172	0.212	0.210	0.057	0.324
N x P Level	1	0.889	0.849	0.007	0.126	0.082	0.923

[a] Degrees of Freedom

significant differences in P concentration were observed among fertilization treatments. The exception, an event that occurred on 24 September 2005, was clearly due to the P fertilizer application made on the previous day. On 23 July 2005, a rainfall event was also simulated 1 day after fertilizer was applied, but in this case, no significant differences in dissolved P concentration were found among the treatments. This finding is likely due to the 15 mm of rainfall that fell overnight between the fertilizer application and the simulated rain event (Table IV). These results are supported by Shuman (*18*) who found light irrigation following fertilizer application can greatly reduce P concentration in runoff from subsequent rainfall events. In general, these findings demonstrate that significant P losses from fertilizer can be expected only in cases where a runoff-causing storm event immediately follows an application of P fertilizer.

Similar to the data collected from natural events, soil P level did not predict runoff P concentrations on any of the 6 simulations dates (Table IV).

Summary and Conclusions

Mass losses of P in runoff were low, as were losses of P associated with fertilizer application. Phosphorus mass losses were 0.05% of fertilizer applied P for both dissolved and total P when corrected for P in the unfertilized control plots. The low mass losses of dissolved P (<0.05 kg ha^{-1} yr^{-1}) can be attributed to the small amount of rainfall that became runoff (0.2%). Application of N or P did not affect the amount of runoff from natural events. However, fertilization

Table IV. Analysis of Variance Table for Dissolved P Concentration in Runoff from Simulated Events

Source of Variation	D.F.[a]	Days after Fertilizer Application					
		18	14	18	1	1	4
		mm of precipitation that fell between fertilizer application and simulated runoff					
		119	89	74	15	0	230
		Simulation date					
		28 July 2004	23 Sept. 2004	27 June 2005	23 July 2005	24 Sept. 2005	27 Sept. 2005
		-------------------------------p-value-------------------------------					
Block	11	0.538	0.089	0.476	0.943	0.484	0.707
Soil P	1	0.119	0.632	0.383	0.634	0.534	0.161
N Level	1	0.581	0.974	0.217	0.128	0.923	0.528
P Level	1	0.528	0.280	0.751	0.321	0.004	0.209
N x P Level	1	0.894	0.171	0.579	0.539	0.362	0.722

[a] Degrees of Freedom

with N and P increased the concentration of dissolved P in the runoff to a similar extent. This suggests that P in turfgrass tissue may be an important P source in runoff from turfgrass areas. The simulated events supported the data observed from the natural events in most cases, indicating that the low cost rainfall simulators used in this study provide useful data for turf runoff studies. Significant differences in runoff P loss occured only when no precipitation fell between a fertilizer application and a simulated runoff event. One difference between the natural and simulated events was that increases in P concentration were not observed with N fertilization for simulated events, as were seen in the natural events. The reason for this is unknown, but is likely related to the differences in runoff from simulated and natural events. If tissue P is indeed an important P source in runoff, then management practices that can reduce these losses should be developed or employed. The results of this study suggest that fertilization of established turfgrass will reduce runoff for this soil type when growth responses to N are similar to those observed in this study. Since increased P concentrations were associated with N or P fertilization, the environmental impacts of turfgrass could be reduced by withholding or limiting N and P fertilizers under these conditions.

Despite Morgan extractable soil P levels ranging from 3.7 to 35.6 mg kg^{-1} at the 0-5 cm depth, soil P levels had no effect on runoff P concentrations or mass losses for both natural and simulated runoff events. This suggests that predictions of runoff P loss from turfgrass areas based on soil P level could be misleading or inaccurate.

106

References

1. USEPA (United States Environmental Protection Agency). 2002. *EPA-F-00-005*. Office of Water, Washington DC.
2. Milesi, C.; Running, S.W.; Elvidge, C.D.; Dietz, J.B.; Tuttle, B.T.; Nemani, R.R. *Env. Management*. **2005**, *36*, 426-438.
3. Linde, D.T.; Watschke, T.L.; Jarrett, A.R.; Borger, J.A. *Agron. J.* **1995**, *87*, 176-182.
4. Easton, Z.M.; Petrovic, A.M. *J Environ Qual.* **2004**, *33*, 645-655.
5. Gross, C.M.; Angle, J.S.; Hill, R.L.; Welterien, M.S. *J. Environ. Qual.* **1991**, *20*, 604-607.
6. Kussow, W.R. Wisconsin Turf Research: *Results of 1996 Studies*. 1996, 14, 1.
7. Sharpley, A.N. *J. Environ. Qual.* **1981**, *10*, 160-165.
8. Cole, J.T.; Baird, J.H.; Basta, N.T.; Huhnke, R.L.; Storm, D.E.; Johnson, G.V.; Payton, M.E.; Smolen, M.D.; Martin, D.L.; Cole, J.C. *J. Environ. Qual.* **1997**, *26*, 1589-1598.
9. Easton, Z.M. *Ph.D. thesis, Cornell University*, Ithaca, NY, 2006
10. Ogden, C.B.; van Es, H.M.; Schindelbeck, R.R. *Soil Sci. Soc. Am. J.* **1997**, 61, 1041-1043.
11. Murphy, J.A., Riley, H.P. *Anal. Chim. Acta.* **1962**, *27*, 31-36.
12. Skogley, C.R.; Sawyer, C.D. In *Turfgrass*; Waddington, D.V. et al., Eds.; Agron. Monogr. 32. ASA-CSSA-SSSA: Madison, WI, 1992, pp 589-614.
13. Morgan, M.F. *Connecticut Agricultural Experimental Station Bulletin* 1941, 1, 450.
14. van Es, H.M.; van Es, C.L., Cassel, D.K. *Soil Sci. Soc. Am. J.* **1989**, *53*, 1178-1183.
15. Easton, Z.M.; Petrovic, A.M.; Lisk, D.J.; Larsson-Kovach, I. *Int. Turfgrass Soc. Res. J.* **2005**, *10*, 121-129.
16. Petrovic, A.M.; Soldat D.; Gruttadaurio, J.; Barlow, J. *Int. Turfgrass Soc. Res. J.* **2005**, *10*, 989-997.
17. Gross, C.M.; Angle, J.S.; Welterien, M.S. *J. Environ. Qual.* **1990**, 19, 663-668.
18. Shuman, L.M. *J. Environ. Qual.* **2002**, *31*, 1710-1715.
19. Linde, D.T., Watschke, T.L. *J. Environ. Qual.* **1997**, *26*, 1248-1254.
20. Soldat, D.J. 2007. *Ph.D. thesis. Cornell University*, Ithaca, NY, 2007.
21. Barten, J.M.; Jahnke, E. *Suburban Lawn Runoff Water Quality in the Twin Cities Metropolitan Area, 1996 and 1997*. Suburban Hennepin Regional Park District, Water Quality Management Division, Maple Plain, MN. 17p.
22. Maguire, R.O.; Chardon, W.J.; Simard, R.R. In *Phosphorus: Agriculture and the Environment*; Sims, J.T.; Sharpley, A.N., eds.; Agron. Monogr. 46; ASA, CSSA, SSSA: Madison, WI, 2005, pp 145-180.

Chapter 7

Influence of Landscape and Percolation on P and K Losses over Four Years

D. M. Park[1], J. L. Cisar[2], J. E. Erickson[3], and G. H. Snyder[4]

[1]Pee Dee Research and Education Center, Clemson University, Florence, SC 29506
[2]Fort Lauderdale Research and Education Center, University of Florida, Fort Lauderdale, FL 33314
[3]University of Florida, Gainesville, FL 32611–0500
[4]Department of Soil and Water Science, University of Florida, Everglades Research and Education Center, Belle Glade, FL 33430

Florida has an intense climate with periods of frequent and heavy rainfall and highly permeable sand soils which have little ability to retain nutrients or water. These factors provide cultural and environmental management challenges for landscape management. Urbanization and land use changes near coastal areas have been shown to degrade water quality. In an effort to reduce fertilizer-based nutrient pollution from urban areas, various programs are promoting alternative landscape materials, which require less input than traditional turfgrass vegetation. Although landscapes that utilize various plant materials that require traditional inputs may conceivably reduce nutrient pollution from urban landscapes, this idea remains unresolved. Here we use a replicated field experiment to show that P and K pollution via leaching was greater on a mixed-species ornamental landscape than a turfgrass monoculture during establishment and routine plant maintenance.

Urban areas with large population concentrations have the potential to strongly impact water resources throughout the world (*1*). Residential and commercial landscapes associated with urban areas have some potential for nutrient losses in surface runoff and leaching due to the predominance of intensely maintained turfgrass areas. After nitrogen (N), phosphorus (P) and potassium (K) are the nutrients applied to turfgrass in the greatest quantity and frequency (2). For example, St. Augustinegrass (*Stenotaphrum secundatum* (Walt.) Kuntze), the most common turfgrass for residential lawns in Florida, is a moderate fertility warm season grass that receives approximately 75-150 kg N ha^{-1} yr^{-1}, 10-34 kg P ha^{-1} yr^{-1}, 35-111 kg K ha^{-1} yr^{-1} when appropriately fertilized (2). Moreover, fertilizer use continues to increase with population growth and residential land use. In 1996, approximately 0.4 billion kg of fertilizer was applied to non-agricultural land use in Florida alone (3).

Several factors have been shown to affect surface runoff and leaching including: 1) vegetation type and density; 2) fertilizer source and rate, 3) frequency and intensity of precipitation event, 4) soil properties and 5) slope (4). Florida soils in residential areas generally are of high sand content, and routinely receive irrigation along with frequent and often intense rainfall. Therefore, conditions exist for rapid percolation and potential for nutrient losses to groundwater, resulting in adverse environmental and economic impacts (5, 6).

As a result of the relatively high maintenance requirements of many traditional landscape materials, a number of authors have proposed the use of alternative plant materials in residential landscapes, which require minimal fertilizer and supplemental irrigation to be maintained in a healthy state (7, 8). While landscapes using these alternative plant materials are generally perceived to require less water and fertilizer inputs, few studies anywhere have examined the loss of fertilizer applied to alternative landscapes. Erickson et al described the effect of landscape model on nitrogen leaching during establishment, and found that a turfgrass monoculture had significantly less leaching than an mixed-species landscape (9). Hipp et al (7) examined the use of resource efficient plants for reducing chemical and nutrient runoff, but came to no generalization with respect to landscape effect on N loss because of conflicts with management practices and other variables. Reinert et al (10) observed less runoff from xeriscapes grown on a silty clay soil, which was attributed to the level of irrigation practiced (antecedent soil moisture). Based on similar resource efficient principles, the Florida Yards and Neighborhoods (FYN) program was established during the 1990s. Partially in response to concerns over non-point source pollution from residential landscapes, the FYN program advocates the use of alternative landscape materials that require fewer inputs and may provide additional environmental benefits over conventional turfgrass lawns (11). Landscapes utilizing the principles of the FYN program are intended to enhance the environment by reducing harmful runoff and providing wildlife habitat. Further considerations include aesthetics, food production, climate control, and resale value (12). Although the FYN landscapes offer a wide variety of potential environmental benefits, a major emphasis is placed on reducing nutrient loading to ground and surface waters since South Florida's watersheds are hydrologically connected (*11*). Furthermore, South Florida wetlands are considered to be P limiting and overall, oligotrophic in nature, and thus P and K losses from residential landscapes may influence the water chemistry and trigger secondary responses such as eutrophication and plant species competitiveness (*13, 14, 15, 16, 17*). However, the FYN program has no P and K runoff and leaching data from FYN landscapes to support their principles. Furthermore, no P and K runoff or leaching data is available for St. Augustinegrass, the predominant turfgrass used for residential landscapes in Florida.

Because non-point source pollution is a pervasive and severe problem in southern Florida as well as throughout the world, a field-scale facility was constructed to monitor runoff and percolate from two contrasting landscape models (a turfgrass monoculture and a mixed-species landscape comprised of native and non-native commercially available groundcovers, shrubs, and trees) managed under current recommendations. The objective of this study was to determine how percolate influences fertilizer P and K concentrations and leaching from the two maturing landscapes.

Materials and Methods

Construction of Experimental Facility

An experiment, consisting of four replications of two yard models: 1) a turfgrass (TG) lawn and 2) a mixed species (MS) landscape, was conducted over a four year period at the University of Florida's Ft. Lauderdale Research and Education Center, Davie, Florida. Each of the eight plots were approximately 5 m wide by 10 m long on a 10% slope subgrade (1.5 m at top sloping to 0.5 m above ground level at the opposite end) consisting of crushed limestone. Test landscapes were hydrologically isolated from each other by 6 mil polyvinyl plastic placed on the bottom (depth of 0.75 m) and sides of each plot to ensure no horizontal water flow into adjacent landscapes (Figure 1). A root zone mix of a mined sand soil (calcitic in origin, initial pH of 8.0-8.5) that was representative of residential soils in South Florida was then applied to both landscapes to a 75 cm depth.

Corrugated drainage pipe was placed at the lower end of each plot under the root zone mix to collect percolate. A hole in the middle of the pipe was retrofitted with a PVC pipe to be used for an outlet (Figure 1). Once the root zone mix was graded, gutters were installed at the lower end of each plot to collect runoff (Figure 1). However, runoff was determined negligible (for example, in year 1 less than 0.175 cm of runoff occurred, 18). See Erickson et al for more details on installation of the facility.

Physical analysis of root zone mix is listed in Table I. Particle size analysis was based on a sample collected before installation, while saturated hydraulic conductivity (K_s), bulk density and total porosity represent means (± s.e.), were based on samples collected from each plot before plant installation (n = 8).

Table I. Physical Properties of the Root Zone Mix Used in the Study

Property	
Particle size (mm)	
Very fine (0.05-0.10)	2.1%
Fine (0.10-0.25)	54.9%
Medium (0.25-0.50)	33.5%
Coarse (0.50-1.0)	8.6%
Very coarse (1.0-2.0)	0.6%
K_S (cm hr^{-1})	59.9±3.3
ρ_b (g cm^{-3})	1.62±0.02
Total porosity (%)	37.8±0.7

SOURCE: Reproduced with permission from reference 17. Copyright 1999 Florida State Horticultural Society.)

Each TG landscape had an in-set sprinkler head system (pop-up sprinklers) to provide overhead irrigation at a rate of 51 mm hr^{-1}. For the first 6 months after the initial fertilization in early February, the MS plots also had an overhead irrigation system using pop-up sprinklers. Thereafter, a micro-injection irrigation system (mist sprayers for a more directed application) was installed and used.

Both the TG, which was St. Augustinegrass (*Stenotaphrum secundatum* Waltz Kuntze cv. 'Floratam') established from sod, as well as the MS were planted in December 1998. The TG landscape was mowed to a 7.5 cm height approximately every two weeks in the wet season and every three weeks in the dry season. Grass clippings were removed for the experimental period except for the second half of the first year of the study. The MS landscape was composed of eleven commercially available species with relatively low maintenance requirements and presumably moderate drought tolerance, seven of which are native to Florida (*21*) (Figure 2 and Table II, adapted from Erickson et al).

Maintenance

Fertilizer and irrigation establishment and maintenance practices were based on recommendations from University of Florida Horticulture personnel and the

Table II. Plant Species and Characteristics Planted in the Mixed Species. RH=River banks, H=Hammocks, S=Swamps, OH=Oak Hammocks, P=Pinelands, F=Flatwoods, D=Dunes, CH=Coastal Hammocks, CD=Cypress Domes, UMF=Upland Mixed Forest, F-BM=Fresh to Slightly Brackish Marshes and EH=Everglade Hammocks

Common Name	Scientific Name	Florida Native	Natural Habitat ‡	#/ Plot	Growth Habit
Liriope	*Liriope muscari* (Dene.) Bailey 'evergreen giant'	N		25	Ground cover
Trailing lantana	*Lantana montevidensis* (K. Spreng.) Briq.	N		3	Ground cover
Dwarf Fakahatchee grass	*Tripsacum floridana* L. 'dwarf'	Y	RB, H, S	3	Ground cover
Coontie	*Zamia pumila* L.	Y	OH, P	15	Ground cover
Dwarf yaupon holly	*Ilex vomitoria* Ait. 'schellings dwarf'	Y	H, S, F, D	7	Small shrub
Firebush	*Hamelia patens* Jacq. 'compacta'	Y	CH	5	Medium Shrub
Thyrallis	*Galphimia gracilis* Cav.	N		4	Medium shrub
Podocarpus	*Podocarpus macrophyllus* (Thunb.) Sweet	N		3	Tree
Simpson's stopper	*Myrcianthes fragrans* (Sw.) McVaugh	Y	CH	3	Large Shrub
Wax myrtle	*Myrica cerifera* L. (small)	Y	H, S, CD,F, UMF, F-BM	1	Tree
Everglades palm	*Acoelorraphe wrightii* (Griseb.& H. Wendl.) H. Wendl. ex Becc.	Y	S, EH	1	Tree
Pink trumpet-tree	*Tabebuia heterophylla* (DC.) Britt.	N		1	Tree

SOURCE: Reproduced with permission from reference 17. Copyright 1999 Florida State Horticultural Society.)

112

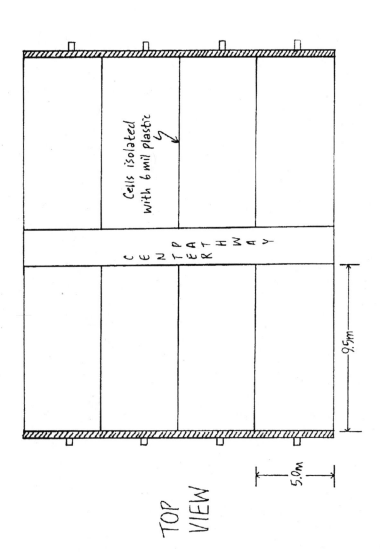

Cells isolated with 6 mil plastic

CENTER PATHWAY

9.5m

5.0m

TOP VIEW

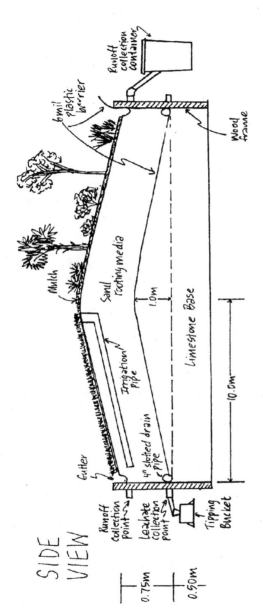

Figure 1. The 20x20m facility constructed to assess surface runoff and leaching from two contrasting landscape models. Eight plots (9.5x5m) were created, which allowed for four replications of each treatment. A plastic barrier (6 mil) provided hydrological isolation for each plot. A gutter system collected any surface runoff, while 10.2-cm slotted pipes drained the percolate. Each landscape was established in 0.75 m of medium-fine sand at a 10% slope. (Reproduced with permission from reference 17. Copyright 1999 Florida State Horticultural Society.)

114

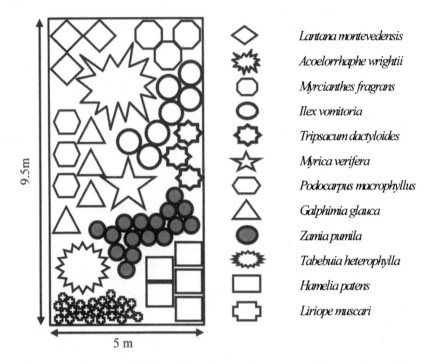

Figure 2. Layout of the mixed-species landscape. The diagram shows the quantity and relative position of all species included within each 9.5x5m plot. The apex of the 10% slope coincides with the top of the diagram.

FYN program (*12, 19*) for the MS landscape, and current recommendations for the TG landscape for home owners (*2, 20*). Each TG landscape had an inset sprinkler head system (pop-up sprinklers) to provide overhead irrigation at a rate of 51 mm hr^{-1}. For the first 6 months after the initial fertilization, the MS plots also had an overhead irrigation system using pop-up sprinklers. Thereafter a micro-injection irrigation system (mist sprayers for a more directed application) was installed and used for the remainder of the experiment. Salt concentrations were low in the municipal irrigation water source with bicarbonate and sodium concentrations equaling 17 mg L^{-1} and 39 mg L^{-1}, respectively.

For the establishment year (YR 1), the TG was managed as according to general requirements for establishing warm season St. Augustinegrass in the subtropics (*2, 20*). Irrigation applied during this time period was equal to 125% potential evapotranspiration (ET$_p$) for both landscapes.

Both landscape treatments were fertilized according to recommendations for residential landscapes in Florida, with a rate of 2.6 kg P ha^{-1} (P$_2$O$_5$) and 18.5 kg K ha^{-1} (K$_2$O) every two months for the TG, and every four months for the MS landscape, using a 26-3-11 granular fertilizer at a rate of 1.4 kg P ha^{-1} and 13.1 kg K ha^{-1}. After two applications to the MS landscape, the fertilizer was changed to a 12-2-14 granular fertilizer, and was applied at 4.1 kg P ha^{-1} (from P$_2$O$_5$) and 10.4 kg K ha^{-1} (from K$_2$O).

During the following two years (YR 2 and YR 3), the TG was irrigated to a soil depth of 2.5 to 5.0 mm after the grass was fertilized and when visual wilting was observed. As in the establishment year, fertilizer was applied every two months to the St. Augustinegrass with the same granular fertilizer. In comparison, the MS landscape was maintained to overcome presumed nutrient deficiencies as judged by visual symptoms by University of Florida horticultural personnel from the FYN program during YR 1. The MS landscape was irrigated thoroughly to rewet the soil profile when visual wilting was observed. The fertilizer was also changed to an 8-4-12 (N-P$_2$O$_5$-K$_2$O) granular fertilizer with P and K from ammonium phosphate and potash respectively. Approximately 50% of the K was encapsulated in a sulfur coated polymer. The fertilizer was applied every two months to the MS landscape at a rate of 9.5 kg P ha^{-1} and 52.9 kg K ha^{-1}.

In the last year (YR 4) of the study period, the TG was irrigated and fertilized on the same basis as the previous two years, while the MS landscape was not fertilized, as it was judged to be nutritionally adequate. However the MS landscape remained irrigated thoroughly to rewet the soil profile (which was determined when percolate was noticed from the drainage pipes) when visual wilting was observed. In total, the TG landscape received 61 kg P ha^{-1} and 421 kg K ha^{-1}, and the MS landscape received 114 kg P ha^{-1} and 634 kg K ha^{-1}. Fertilizer and irrigation maintenance is summarized in Table III.

Water Sample Collection and Chemical Determination

Initially, soil water (leachate) samples, along with percolate flow measurements, were taken at least once daily from a slotted drainage pipe placed across the lower edge of each plot, which drained the percolate for the entire plot (Figure 1). After approximately 6 months, percolate samples were collected on a Monday-Wednesday-Friday schedule, and after rain events. Data loggers (Campbell Scientific® CR10X) recorded percolate quantity from tipping bucket flow gauges (Unidata America®-Model 6406H), calibrated to tip upon collection of 125 ml, connected to the PVC outlet located in each plot. Percolate samples collected were immediately acidified with sulfuric acid upon collection and refrigerated at 4 °C until analysis. The samples were analyzed for P and K by atomic emission spectroscopy (Varion ICP-OES®, Varian Inc., Palo Alto, California) at the University of Florida Analytical Research Laboratory in Gainesville, Florida. The quantity of nutrients leached (in kg ha^{-1}) was calculated by multiplying the concentration (in mg L^{-1}) of each nutrient found in the daily composite sample by the volume of percolate measured for the respective 24 hour period. Annual leaching is the sum of all daily leaching over each year. Monthly and annual concentrations are flow weighted means, and

Table III. Fertilization and Irrigation Schedule for the Turfgrass and Mixed-Species Landscapes over the Four Year Study Period

Landscape Model	Maintenance Schedule
Turfgrass	Fertilizer: 0- 48 MAI[a]: 26-3-11 applied every 2 months at a rate of 50 kg N ha^{-1}. Irrigation: 0-12 MAI: Overhead perimeter sprinklers applied 125% ETp. 12-48 MAI: Overhead perimeter sprinklers applied 2.5-5.0 mm after fertilization and when visual wilt was observed.
Mixed Species	Fertilizer: 0 and 4 MAI: 26-3-11 applied at a rate of 37.5 kg N ha^{-1}. 8 MAI: 12-3-4 applied at a rate of 37.5 kg N ha^{-1}. 12 MAI: 8-4-12, applied every 2 months at a rate of 40.0 kg N ha^{-1} 36 MAI: No fertilizer applied. Irrigation: 0-6 MAI: Overhead perimeter sprinklers applied 125% ETp. 6-12 MAI: Microjet sprinklers applied 125% ETp. 12-24 MAI: 2.5-5.0 mm after fertilization and then $\geq$ 30.5 mm when visual wilt was observed. 24-48 MAI: $\geq$ 30.5 mm when visual wilt was observed.

[a] MAI=months after initiation.

were calculated as total leaching over the period divided by the total percolate during the period.

Both P and K concentration data were lost during October 1999 through January 2000 due to equipment failure. To include this time frame within the annual leaching totals, the average nutrient concentrations found in percolate from February to October 1999 was multiplied by total drainage over the time period for each plot. While this may not represent exactly the leaching during this time period, it does fall within the establishment year. Furthermore, it represents a worst case scenario for P since that is when P leaching was greatest. Additionally, percolate data were lost from February and March 2000 due to data loggers malfunction. No leaching data is reported for this time. Runoff was monitored, but not included into the equation, since runoff was observed only on two rainfall events in June and November 1999, both of which had negligible runoff volumes (9).

Rainfall, ET_p (modified Penman-Montieth) and temperature at six meters above ground were downloaded through the Florida Automated Weather Network (FAWN) from a weather station located in a grass field approximately 100 m from the experimental facility.

Statistical Analysis

The experimental design for this study was a completely randomized design with a single factor, landscape model. The design included two treatments and four replications. Mean treatment effects were determined for annual average nutrient concentrations and total leaching losses according to the four, 1 year experiments. To study the dynamics of P and K concentrations within percolate and leaching losses from the landscapes, correlation between percolate and P and K concentrations and leaching may help to understand how each landscape utilized the nutrients. More specifically, low correlations for percolate with concentrations and leaching may result from the nutrients' fate other than by loss to percolate. High correlations of percolate to nutrient concentrations and leaching may result from the plants in the landscape obtaining nutrient requirement. For each year, monthly percolate was correlated with monthly P and K average concentrations and total leaching. Statistically significant treatment effects on P and K annual average concentrations and total leaching losses were identified using procedures for ANOVA.

Results and Discussion

Rainfall, Irrigation and Percolate

Rainfall for years 1 through 4 reflected the typical pattern of rainfall for south Florida (Figure 3, Table IV) including a wet season from approximately May - October and a dry season approximately from November - April. During the establishment first year, the plantings were grown during a dry season that experienced less than normal precipitation, followed by a wet season that experienced two large single storm events providing an overall precipitation of over 2000 mm (Table IV). The second year had an even more severe dry season, followed by a drier than normal wet season with rain events that were less intense and more evenly spaced for a total of precipitation nearly 66% of that measured during the first year. The third year had a weather pattern more closely associated with longterm weather in south Florida, while the fourth year had less rainfall.

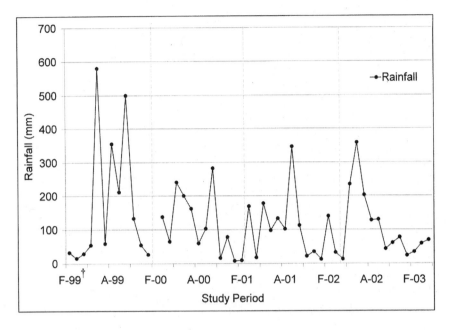

Figure 3. Monthly rainfall (mm over the four year study period
(Feb. 1999 – April 2003). Vertical hatched lines delineate years.
[†] F=February and A=August for corresponding year.

Overall, 1408 and 1404 mm of irrigation were applied over the 4 year experimental period to the TG and MS landscapes, respectively. Frequent irrigation required to ensure plant establishment in the beginning of the 1988 dry season resulted in applying the largest amounts of irrigation and having the greatest percolate for the first year (Table IV). The TG landscape required significantly more irrigation than the MS landscape during the first two years of the experiment (Table IV). Irrigation requirements increased in years 3 and 4 for the MS landscape, with a significantly greater amount of irrigation applied to the MS landscape than the TG landscape in year 4 (Table IV). This can be attributed to the maintenance of irrigating both landscapes only when visual wilt was present. The amount of water applied to the TG during stress was minimal since less water was required for the grass to be revived. While irrigation was needed on a more frequent basis for the TG landscape, the quantity applied per application was less than the minimum of 30.5 mm applied per irrigation in year 2 to replenish the entire root zone profile, and also less than the up to 61 mm per irrigation in year 4 needed for the MS landscape.

Table IV. Rainfall and Irrigation Applied and Drainage Percolate
Measured (mm yr^{-1}) for the Turfgrass and Mixed Species Landscapes over
the Four Year Study Period (February 1999 – April 2003)

| | | Irrigation | | Percolate | |
| | | | Mixed | | Mixed |
Year (Dates)	Rainfall	Turfgrass	Species	Turfgrass	Species
1 (Feb 1999 – Feb 2000)	2054	964[a]	881	2080	2240
2 (Jun 2000 – May 2001)	1394	233[a]	104	804[a]	555
3 (Jun 2001 – May 2002)	1529	102	165	681	633
4 (Jun 2002 – Apr 2003)	1423	109[a]	254	655[a]	514

[a] Indicates significant differences for landscape model during the corresponding year at P<0.05.

Percolate was primarily driven by large rain events, and in the case of the MS landscape, applications of deep, infrequent irrigation. Interestingly, one rain event in the first year resulted in nearly as much percolate as the entire second year (Figure 4). In both landscapes, percolate decreased after year 1, with the TG landscape having significantly greater percolate than the MS landscape in year 2 (804 and 655 mm yr^{-1}, respectively, P<0.05) and year 4 (655 and 514 mm yr^{-1}, respectively, P<0.05, Table IV). There was minimal change in the total yearly percolate from the MS landscape after year 1, and after year 2, for the TG landscape (Table IV, Figure 4).

Average [P] and Total P Leaching

Average [P] in rainfall was 0.053 +/- 0.024 mg P L^{-1}, contributing a total of 3.2 kg P ha^{-1} over the four year study period. Average [P] in the irrigation source was 0.017 +/- 0.017 mg P L^{-1}, with a total of 0.24 kg P ha^{-1} applied to each landscape model over the four year study period.

Regardless of landscape model, [P] was the highest during the establishment year (YR 1), when both landscapes were routinely irrigated and after fertilizer applications (Figure 5). However, mean [P] found in percolate from the MS landscape was twice the mean [P] found in percolate from the TG landscape (1.21 and 0.59 mg l^{-1} for MS and TG, respectively, P=0.0205). The decline in [P] for both landscapes in YR 2 might have been influenced by a growing root system, the change in irrigation with a "water only upon visual stress" irrigation schedule, and also in regard to the MS landscape, a change to a slow release P source (Figure 5). Mean monthly [P] found in percolate from the TG landscape for the remainder of the study period (YRS 3 and 4) remained below 0.50 mg L^{-1} except for January 2002 and during the 2002 wet season.

120

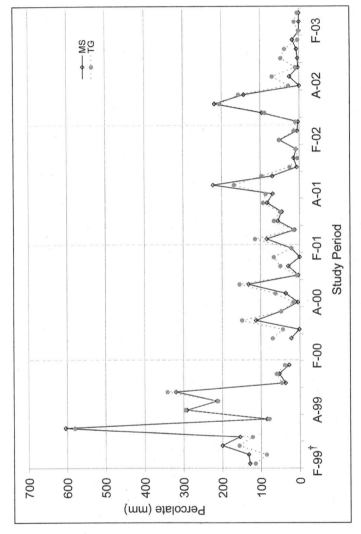

Figure 4. Monthly percolate (mm) for the mixed-species (MS) and turfgrass (TG) landscapes over the four year study period (Feb. 1999 – April 2003). Vertical hatched lines delineate years. † F=February and A=August for corresponding year.

Mixed Species landscape [P] declined in YR 4, however [P] spiked in some months (Figure 5).

Phosphorus leaching was greatest from both landscapes during the establishment period (YR 1) with leaching peaking in June as a result of numerous rain events (Figures 3 and 6). For the rest of the study period, P leaching was generally below 4 mg L^{-1} mo^{-1} for both landscapes, with no significant differences in total P leached for YRS 2 and 4 (Table V). Phosphorus leaching from the two landscapes was similar, even though the P source for the MS landscape was applied as a slow-release form in YR 2 and no fertilizer was applied to the MS landscape in YR 4. Phosphorus leaching from the MS landscape during YR 3 was greatest during the wet season, and rain most likely caused significantly greater P leaching from the MS landscape compared to the TG landscape (Table V and Figure 6). The slow release P applied during YR 3 may have influenced P concentrations and leaching in the earlier part of YR 4, when no fertilization was applied to the mixed species landscape. Perhaps the breakdown of organic matter below ground (roots and microbial fraction) and above ground (leaves and mulch) also contributed to soluble P that was leached from the MS during the precipitation events. However, the presence of P concentrations in leachate from the MS not only during large precipitation events provided evidence that considerable quantities of P were leached beyond the root zone. In summary, large precipitation events influenced P leaching. For example, the 165mm of rainfall occurring two days after a fertilization event (May 17th) resulted in the greatest P loading for YR 4 for both landscapes (Figures 3 and 6).

Over the four year study period, totals of 113.5 and 64.4 kg P ha^{-1}, respectively, were applied to the MS and TG landscapes by irrigation, rainfall and fertilizers. The MS landscape leached 41 kg P ha^{-1} over the period, or approximately 36% of P inputs. This is comparison to the TG landscape which leached 23 kg P ha^{-1}, or 36% of total P inputs Thus, landscape model did not affect the percentage of P leached. However, because of the differing requirements of the two landscapes, the MS landscape leached more P than did the TG landscape.

Correlation of Percolate with [P] and P Leaching

Monthly [P] in percolate from the MS landscape was not correlated with percolate during YRS 1,2, and 4 (P>0.1000, Table VI). Only in YR 3 was [P] correlated to percolate (R=0.60, P<0.05, Table VI). However for all years, percolate was significantly correlated to P leaching (P<0.05, Table VI). Perhaps this reflects the fertilization schedule over the four year period. We postulate that during the establishment year (YR 1), and the first year that the rate of fertilizer was increased due to observations of nutrient deficiencies (YR 2),

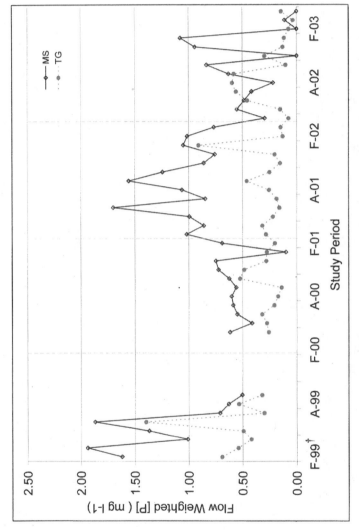

Figure 5. Monthly flow weighted [P] (mg L^{-1}) for the mixed- species (MS) and turfgrass (TG) landscapes over the four year study period (Feb. 1999 – April 2003). Vertical hatched lines delineate years. [†] F=February and A=August for corresponding year.

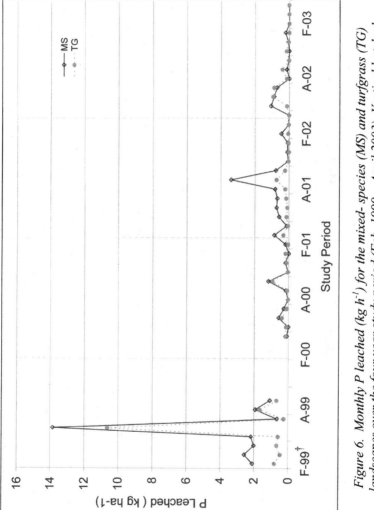

Figure 6. Monthly P leached (kg h⁻¹) for the mixed-species (MS) and turfgrass (TG) landscapes over the four year study period (Feb. 1999 – April 2003). Vertical hatched lines delineate years. † F=February and A=August for corresponding year.

Table V. Comparison of Total Annual P and K Leaching (kg ha^{-1}) for the Turfgrass and Mixed Species Landscapes over the Four Year Study Period (February 1999 – April 2003)

	Phosphorus		Potassium	
		Mixed-		Mixed-
Year (Dates)	Turfgrass	Species	Turfgrass	Species
1 (Feb 1999 – Feb 2000)	15.74[a]	26.36	2.37[a]	10.15
2 (Mar 2000 – Mar 2001)	2.61	3.67	83.04	83.36
3 (Apr 2001 – Mar 2002)	1.68[a]	7.85	60.68[a]	138.17
4 (May 2002 – Apr 2003)	2.78	3.16	22.16[a]	40.89

[a] Indicates significant differences for landscape model during the corresponding year at $P < 0.05$.

perhaps [P] in soil solution was not efficiently removed from the soil solution by plant roots for the plant to utilize. It may have thus been left in the soil profile and subsequently prone to leaching during percolate events. During the second year of correcting nutrient deficiencies by applying the higher rate of fertilizer (YR 3), the MS landscape had matured and overcome nutrient deficiencies, leaving the unused balance of added P in the soil and subsequently vulnerable to leaching events. In YR 4, when no fertilizer was applied, the MS landscape utilized P reserves in the soil (presumably by fertilizer added P not leached because of dry season conditions) and/or recycled P to maintain the appropriate levels within the MS plants, once again depleting available P. It is possible that during YRS 1, 2 and 4, the low correlations may also be influenced by soil moisture, microbial activity or complicated by the accumulated organic material within the soil from decaying leaf litter. However, these P fate pathways were not examined in this study.

Percolate was significantly correlated with P leaching from the TG landscape for all four years of the study (Table VI). Phosphorus concentrations were significantly correlated with percolate from the TG landscape in YR 1. During this establishment year, the TG landscape was frequently irrigated and perhaps because of the establishing root system, most of the applied P was lost with percolate. Another factor might have been that the likelihood of any microbial community associated with the root zone mix was minimal since the facility was newly constructed. After the first year, the TG landscape was irrigated only upon visual wilt. During these next two years, percolate was not correlated to [P] (Table VI). The lack of correlation between percolate and [P], the high correlation between P leached and percolate, coupled with no observations of nutrient deficiencies suggests that the soil moisture was dictating [P]. In YR 4, percolate was once again significantly correlated with [P] in TG and might have been attributed to the more frequent irrigation in YR 4 in response to visually observing wilt during dry season months.

Table VI. Annual Correlation (R) of Percolate with Flow Weighted P and K Averages and Total P and K Leached for each Landscape Using Monthly Averages and Totals

	Mixed Species				Turfgrass			
	YR 1	YR 2	YR 3	YR 4	YR 1	YR 2	YR 3	YR 4
[P]	0.26	0.34	0.60[a]	0.14	0.84[a]	0.38	-0.02	0.62[a]
[K]	-0.28	0.30	0.08	0.14	-0.02	0.01	0.67[a]	0.30
P leached	0.92[a]	0.94[a]	0.98[a]	0.89[a]	0.94[a]	0.85[a]	0.92[a]	0.95[a]
K leached	0.56	0.84[a]	0.96[a]	0.58[a]	0.85[a]	0.79[a]	0.94[a]	0.89[a]

[a] Indicates significant correlation at P<0.05.

Average [K] and Total K Leaching

Average [K] in rainfall was 2.93 +/- 0.581 mg K L^{-1}, contributing a total of 187.5 kg K ha^{-1} over the four year study period. Average [K] in irrigation source was 3.02 +/- 0.031 mg K L^{-1}, with a total of 42.4 and 42.5 kg K ha^{-1} applied to MS and TG respectively over the four year study period.

Potassium concentrations peaked after landscape installation for both landscapes in YR 1, with initial [K] from the MS landscape being three times greater than [K] from the TG landscape (Figure 7). Even though the K was readily soluble for the TG treatment and slow release for the MS landscape for the second year, [K] from both landscapes rose from less than 5 mg L^{-1} at the beginning of the year, to over 15 and 35 mg L^{-1} for TG and MS landscapes, respectively, by the end of the experimental year (Figure 7). While more pronounced from the MS landscape, [K] found in percolate from both landscapes were strongly influenced by seasonal rainfall patterns (Figures 3 and 7). However, the highest [K] in percolate from the MS landscape was during dry season months when the landscape had to be irrigated due to visual signs of stress (Figure 7). Lower [K] from percolate from both landscapes occurred during YR 4 in comparison to YR 3 (Figure 7). The mean [K] in percolate from the MS and TG landscapes during YR 3 were 23.0 and 7.0 mg L^{-1}, respectively. This is in comparison to the 2.2 and 3.1 mg L^{-1} mean [K] in YR 4 from the MS and TG landscapes, respectively.

Potassium leaching followed similar trends to [K], with K leaching greatest at the onset of the experiment and in YR 3 (Figures 7 and 8). As with [K] in YR 1, the magnitude of difference in K leaching between the two landscapes was significant, with 10.15 kg ha^{-1} of K leached from the MS landscape compared to the 2.37 kg ha^{-1} of K leached from the TG landscape (Table V, Figure 8). In YR 2, K leaching was similar for both landscapes, presumably due to the differences in percolation and to the amounts of applied K since the MS had more K applied (Tables IV and V, Figures 4 and 8). Potassium leaching during YR 3 was greater from the MS landscape compared to the TG landscape (138.17 and 60.68 kg ha^{-1}

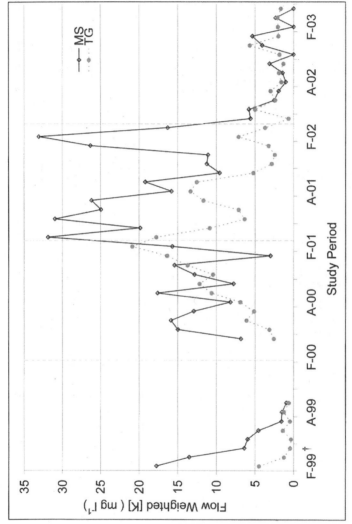

Figure 7. Monthly flow weighted [K] (mg L⁻¹) for the mixed- species (MS) and turfgrass (TG) landscapes over the four year study period (Feb. 1999 – April 2003). Vertical hatched lines delineate years. † F=February and A=August for corresponding year.

for MS and TG, respectively, Table V). Additionally, total K leaching from the MS landscape increased from YR2 to YR3, while K leaching decreased from the TG landscape (Table V, Figure 8). Perhaps similar to P leaching, the increase in K leaching from the MS landscape can be attributed to the plants having received a higher K fertilization and with plant requirements fulfilled, more K was left in the soil profile to be leached. As with [K], less K leaching from the MS and TG landscapes occurred during YR 4 in comparison to YR 3 (Table VI, Figure 8). During YR 4, no fertilizer was applied to the MS landscape, however K leaching peaks continued to be observed during the onset of YR 4, and began during the dry season when the landscape was irrigated. An annual total of 40.89 kg ha^{-1} leached from the MS landscape was less than half of what was leached in the first year that K applied to the MS was increased (YR2), and suggests that K leaching was influenced by the presence of K available in the soil solution. Since K was applied to the MS landscape as a slow release form in YR 2 and YR 3, K from previous fertilizations may also explain the large peak in K concentrations and leaching from the MS landscape in May of YR 4. Over the four year study period, 864.2 and 650.5 kg K ha^{-1} was applied through rainfall, irrigation and fertilizer to the MS and TG landscapes, respectively. In total, 351.1 and 184.7 kg K ha^{-1} was leached from the MS and TG landscapes, respectively, representing 41% and 28% (for MS and TG landscapes respectively) of what was applied from irrigation, rainfall and fertilizers.

Correlation of Percolate with [K] and K Leaching

Percolate was not correlated with [K] from the MS landscape during the study period (Table VI) suggesting that the MS landscape was utilizing the applied K. However, after YR 1, percolate was significantly correlated with K leaching for the remainder of the study period (Table VI). This correlation was highest during YR 2 and YR 3 (R=0.84 and 0.96 for YR 2 and YR 3, respectively) when higher rates of K were applied to overcome nutrient deficiencies. Perhaps during this time, the higher rate of K applied was more than what plants in the MS landscape required. Percolate was not significantly correlated to K leaching during the establishment year (YR 1) even though that year received more irrigation. The lack of correlation was most likely due to the plants within the MS becoming established (Table VI).

In the first two years of the study, the very low correlation of percolate from the TG landscape to [K] suggests that [K] was governed by processes other than percolate. Percolate from the TG landscape was significantly correlated to TG percolate [K] during YR 3 but not in YR 4 (R=0.67, P=0.0167 and 0.30, P=0.5914 for YR 3 and YR 4, respectively). As mentioned earlier, YR 3 [K] over the entire year was higher compared to the previous and past years (YR 2 and YR 4, Figure 7). As with P, K leaching during all four years was highly correlated with percolate (Table VI), suggesting that regardless of other fates of

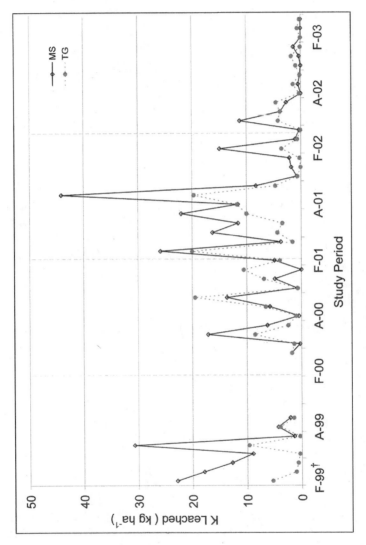

Figure 8. Monthly K leached (kg h⁻¹) for the mixed- species (MS) and turfgrass (TG) landscapes over the four year study period (Feb. 1999 – April 2003). Vertical hatched lines delineate years. † F=February and A=August for corresponding year.

K, soil moisture was most likely the driving factor in whether the nutrient was leached or not.

Conclusions

Similar irrigation requirements for the TG landscapes during the 3 years after the establishment year period reflected the similar annual TG ET reported as between 822-948 mm of water per year. However, as this experiment progressed, the amount of irrigation required for the MS landscape increased. Thus, even with prudent use of irrigation adjusted to meet plant demands (based on visual plant stress) water use by the MS landscape increased and it probably will until the landscape reaches a maximum canopy. Increased canopy provides a greater surface area for water loss. Because plant material was only irrigated upon visual stress, some of the plant materials declined, requiring more water to recover from subsequent periods of stress. In comparison, the TG landscape recovered from water stress after minimal irrigation. It was evident from the past two years that deep infrequent irrigations were not acceptable for all the plants in this maturing MS landscape. Different irrigation scheduling for MS landscapes and or different groupings of MS plant materials (matched by growth habit or water use) would be of interest for future research.

Since loading response was amplified in the MS landscape during large precipitation events, infrequent but deep irrigation applications may not be completely suitable for reducing potential leaching. Perhaps nutrients released from the mineralization of decomposing organic matter from below and above ground sources was later leached with precipitation events. In order to minimize large nutrient concentrations exposed to precipitation events, different irrigation regimes and fertilization rates and frequencies should be considered for the MS landscape.

This study tested whether percolate alone was the main reason for [P] and [K] and P and K leaching from two landscapes. The correlations of percolate and P and K leaching for both landscapes, as well the change in correlations of percolate to [P] and [K], suggest that fertilizer application alone may not be the only driving force behind P and K losses to water sources. Rather, as shown multiple times through out the four years of this study, results indicated that nutrient loss was more closely linked to precipitation and subsequent percolate events. Nutrient loading occurred when 1) high concentrations were observed during periods of low precipitation, or 2) with lower concentrations during periods of greater precipitation. This may also explain the lack of correlation between percolate and [P] and [K], since the correlations were over a year period.

Nutrient deficiencies on some plant materials (i.e. Everglades Palm and Coontie) were becoming more apparent towards the middle to end of YR 4. It is possible that if the fertilization curtailment on MS landscape continues, plant materials may further decline. It is suggested that other plant materials be investigated and or different fertilization schedules be investigated for this landscape.

Although the turfgrass treatment had lower nutrient concentrations in the leachate and less nutrient leaching than the MS landscape, nutrient concentrations above those considered having environmental impact were observed for both landscape models. Research is needed to identify management strategies that result in reduced nutrient losses from any type of landscape. Research focusing on lower rates of P and K, alternative P and K sources, application frequency intervals during both establishment and through maintenance, soil modification, and identification of suitable plant materials with lower P and K requirements are suggested.

Acknowledgements

We thank Karen Williams, Kevin Wise and Scott Park for technical support in the field and David Rich for laboratory assistance. The authors would also like to acknowledge the anonymous reviewers for their comments. This work was supported by the Florida Department of Environmental Protection, Sarasota Bay National Estuary Program, and the University of Florida Agricultural Experiment Station.

References

1. Niedzialkowski, D. and D. Athayde. In *Perspectives on Nonpoint Source Pollution, Proc. National Conf., May 19-20, 1985*, Kansas City, MO, 437-441.
2. Cisar, J. L., G. H. Snyder, and P. Nkedi-Kizza. *Maintaining Quality Turfgrass with Minimal Nitrogen Leaching.* Univ. of Fla., Inst. of Food and Agr. Sci. Bul. 273. 1991, 11 pp.
3. USEPA. *Background Report on Fertilizer Use, Contaminants and Regulations.* EPA 747-R-98-003. U.S. Environmental Protection Agency, Washington, D.C., 1999.
4. Coale, F.J., F.T. Izuno, and A.B. Bottcher. *J. Environ. Qual.* **1994**, *23*, 121-126.
5. Correll, D. L. *J. Environ. Qual.* **1998**, *27*, 261-266.
6. Hipp, B., S. Alexander, and T. Knowles. *Water Sci. Tech. 28* (No. 3-5). **1993**, 205-213.
7. Brady, N. C. *The Nature and Properties of Soils.* Macmillan Publishing Company, New York. Tenth ed. 1990, 621 pp.
8. Sacamano, C. M. and W. D. Jones. *Univ. AZ Coop. Ext. Serv. Bull. A-82.* 1975, 40 pp.

9. Erickson, J.E., J.L. Cisar, J.C. Volin, and G.H. Snyder. *Crop Sci.* **2001**, 41:1889-1895.
10. Reinert, J. A., S. J. Maranz, B. Hipp, and M. C. Engelke. In *The Pollution Solution: Be Water Wise Proceedings.* Texas A&M Univ. Dallas, Texas. 1997, 7-31.
11. Best, C. H. *Proc. Fla. State Hort. Soc.* 1994, *107*:368-370.
12. Garner, A., J. Stevely, H. Smith, M. Hoppe, T. Floyd, and P. Hinchcliff. *U. of Florida Institute of Food and Ag. Sciences Bulletin 295.*1996, 56 pp.
13. Davis, S.M. *In Everglades: The Ecosystem and its Restoration*, S.M. Davis and J.C. Ogden (eds.) St. Lucie Press, Delray Beach, FL. 1994. p. 419-444,
14. Noe, G.B, D.L. Childers, and R.D. Jones. *Ecosystems.* **2001**, 4:603-624.
15. Tilman, E.A., D. Tilman, M.J. Crawley, and A.E. Johnston. *Ecol. Applications.* **1999**, 103-111.
16. Ewel, K.C., and H.T. Odum. Eds. *Cypress Swamps.* Univ. Press. of Fl. Gainesville. 1984, 472 pp.
17. Koch, M.S., and K.R. Reddy. *Soil Sci. Soc. Am. J.* **1992**, 56:1492–1499. Erickson, J.E., J.C. Volin, J.L. Cisar, and G.H. Snyder. *Proc. Fla. State Hort. Soc.* 1999, 112:266–269.
18. Gilman, E. F. and R. J. Black. *Your Florida Guide to Shrubs.* University Press of Florida, Gainesville, FL. 1999, 116 pp.
19. Yeager, T.H., and E.F. Gilman. *Univ. of Fla. Coop. Ext. Serv., Circ. 948.* Univ. of Florida, Gainesville, FL. 1991.
20. Wunderlin, R. P. *Guide to the Vascular Plants of Florida.* University of Florida Press, Gainesville, Florida, USA. 1998.

Chapter 8

Management Practices That Reduce Runoff Transport of Nutrients and Pesticides from Turfgrass

Gregory E. Bell[1] and Justin Q. Moss[2]

[1]Department of Horticulture and Landscape Architecture,
Oklahoma State University, Stillwater, OK 74078
[2]Sheridan Research and Extension Center, University of Wyoming,
Sheridan, WY 82801

Research in crop production and turfgrass has identified grass swards and grass buffer strips as impediments to the transport of nutrients and pesticides in runoff. Dense grass swards have unique characteristics that encourage water to infiltrate soil and impede and filter runoff. Research has also demonstrated that the runoff-reduction characteristics that naturally occur in a grass sward are not sufficient to prevent the substantial runoff caused by major storm events. Urban turfgrasses are usually managed to provide relatively high aesthetic and functional value. Maintenance applications of fertilizer and pesticides required to satisfy consumer expectations, when followed by major storm events, can result in unsatisfactory product transport to surface water features. Practicing runoff reduction management strategies can significantly reduce the occurrence of nutrient and pesticide runoff from turfgrass sites.

Surface runoff from turf occurs either when the precipitation rate exceeds the surface infiltration rate of a turf/soil system and free water accumulates on the system surface and/or the soil becomes saturated. Before and during a precipitation event, many factors combine to affect the volume of runoff that will occur for any given precipitation rate on any given turfgrass stand. These factors include antecedent soil moisture (*1*) relevant soil physical properties, (*2*), turfgrass species (*3*), turf density (*4*) and land slope at the site (*5*). Factors such as leaf orientation, thatch thickness and composition, and turfgrass health and rigidity may also affect the occurrence and volume of runoff. Once surface runoff begins, flow may be channeled through natural and artificial drainage into surface water features such as lakes, ponds, and streams. Normally, surface runoff from turf has little environmental impact (*6*). However, because maintenance applications of nutrents and pesticides are required to maintain color and density at commercially or socially acceptable levels, there is a danger that some portion of a recent nutrient or pesticide application may combine with surface water runoff and flow into adjacent water features. If appropriate management techniques are practiced before, during and after nutrients or pesticides are applied, nearly all of this chemical runoff can be prevented.

Runoff Transport of Nutrients and Pesticides

Nutrient Runoff

An important environmental hazard caused by nutrient runoff is eutrophication (*7*). Low levels of nitrogen (N), mostly in the form of nitrate (NO_3^-), and dissolved reactive phosphorus (DRP), primarily as forms of phosphate ($H_2PO_4^-$, HPO_4^{2-}, and PO_4^{3-}), can cause algal blooms resulting in a loss of oxygen in surface water. This process is called eutrophication. Eutrophication is responsible for the "dead zones" in the Mississippi Delta and the Chesapeake Bay as well as numerous lakes and other water features throughout the world. At least one state, Minnesota, has passed legislation that restricts the application of phosphorus fertilizer to turfgrass (*8*). Nitrate in surface water at concentrations as low as 1 mg L^{-1} may lead to eutrophication (*9*). High NO_3^- levels in drinking water are also a human health hazard. The United States Environmental Protection Agency (USEPA) has established a drinking water standard of 10 mg L^{-1} for NO_3-N (*10*).

Generally, about 99% of the phosphorus (P) in soils is unavailable for plant growth because most is not in a plant-available form (*11*). The availability of P depends on soil pH; the amount of soluble iron (Fe), aluminum (Al), and manganese (Mn) in the soil; the amount of Fe-, Al-, and Mn-containing minerals in the soil; the amount of calcium (Ca) and Ca minerals in the soil; the amount and rate of decomposition of organic matter; and the microorganism activity in the soil (*11*). Fertilizers are thus important as a source of plant available P. Most inorganic fertilizers, however, are highly soluble and if not properly applied, increase the risk of P loss to surface runoff (*12, 13*). Phosphate-P can contribute to eutrophication at concentrations as low as 25 μg L^{-1} (*14*) and is typically the limiting factor for eutrophication of surface water (*15*).

Nutrient transport in surface runoff is affected by factors including rainfall or irrigation amount, intensity and duration of rainfall or irrigation, soil moisture, soil texture, slope, fertilizer application rate, and fertilizer formulation (*5*).

Pesticide Runoff

Pesticide loss from turf depends on factors such as pesticide chemical properties, soil type, turf species, thatch, application timing and weather conditions (*5, 16*). Pesticides may be transported to surface water through runoff or eroded sediment. Cohen et al (*6*) analyzed water quality data from eighteen studies on golf courses in the United States and one study in Canada. Thirty-one pesticide chemicals were detected in surface waters, nine exceeded maximum allowable concentrations for aquatic organisms, and five exceeded maximum contaminant levels for drinking water. The average concentration of the pesticides ranged from 0.07 to 6.8 μg L^{-1}. Transport of pesticides such as 2,4-D [(2,4-dichlorophenoxy) acetic acid], dicamba (3,6-dichloro-2-methylphenoxy-benzoic acid), and mecoprop [(±)-2-(4-chloro-2-methylphenoxy)-propanoic acid] in runoff from turfgrass can be significant if the soil is saturated and rainfall duration and intensity are high (*17*). Smith and Bridges (*17*) for instance, found that 9, 14, and 13% of the applied 2,4-D, dicamba, and mecoprop, respectively, was lost to runoff from hybrid bermudagrass [*Cynodon dactylon* (L.) x *C. transvaalensis* Burtt-Davy] during four simulated rainfall events over an 8-day period. Researchers have concluded that the greatest mass and concentration of pesticides in runoff from a turf area occur during the first significant runoff event after pesticide application (*1, 17, 18*) and that the amount of pesticide loss is primarily related to its solubility (*19*).

Role of Turf as a Deterrent to Runoff

Krenitsky et al (*20*) compared natural and manmade erosion control materials. They found that tall fescue (*Festuca arundinacea* Schreb.) sod was an effective material for delaying the start of runoff and decreasing total runoff volume. Gross et al (*21, 22*) studied nutrient and sediment losses from turf and found that turfgrass alone (without buffers) effectively reduced nutrient and sediment losses compared with bare or sparsely vegetated soil. Linde et al (*23*) found that sediments in runoff were low even after vertical mowing of creeping bentgrass (*Agrostis stolonifera* L.) and perennial ryegrass (*Lolium perenne* L.) turf. Wauchope et al (*24*) investigated pesticide runoff from bare soil plots compared with grassed plots and determined that the bare plots required one third less precipitation to produce the same amount of runoff and yielded twice as much sediment as the grassed plots.

Harrison et al (*25*) determined nutrient and pesticide concentrations in runoff from sodded Kentucky bluegrass (*Poa pratensis* L.). Plots were fertilized with N, P and K in a typical maintenance program for golf course turf in the northeast United States. Irrigation at rates of 75 mm h^{-1} and 150 mm h^{-1} for one hour was applied one week prior to, and two days following, fertilizer applications. The researchers reported that N concentrations were as high as 5 mg L^{-1} and dissolved P concentrations were as high as 6 mg L^{-1}. Both N and P nutrient concentrations were above those that can cause eutrophication of surface waters. The researchers concluded, however, that under the conditions studied, nutrient runoff from established turfgrass areas was not affected by establishment method.

Gross et al (*21*) studied nutrient and sediment loss from sodded tall fescue and Kentucky bluegrass plots. The plots were sodded on land that was previously cropped to tobacco (*Nicotiana tabacum* L.). Land slope at the site was 5 to 7%. Plots were fertilized with either urea dissolved in water as a liquid application or urea as a granular application at a rate of 220 kg N ha^{-1} yr^{-1}. Control plots were not fertilized. Nutrient and sediment losses were low for all replications. The researchers concluded that nutrient and sediment runoff from turfgrass areas was low, especially when compared with the previously cropped tobacco runoff study. Gross et al (*22*) studied runoff and sediment losses from tall fescue stands of various density under simulated rainfall conditions. Plots were established at seeding rates of 0, 98, 244, 390, and 488 kg ha^{-1} in September 1986. Simulated rainfall was applied at intensities of 76, 94 and 130 mm hr^{-1} in June 1987. The highest runoff volume was observed from the non-seeded plots at each rainfall intensity. Runoff volume was not statistically different among the 98, 244, 390, and 488 kg ha^{-1} seeding rates. The researchers also recorded visual quality, density, and tiller counts. They concluded that even low-density turfgrass stands can significantly reduce surface water runoff from well-maintained turfgrass areas.

Kauffman III and Watschke (*26*) studied phosphorus and sediment runoff from creeping bentgrass and perennial ryegrass following core aerification They concluded that the DRP concentrations found in the runoff and the minimal soil erosion that occurred should not be considered a serious threat to surface waters.

When turfgrass is healthy and dense, it is an effective deterrent to offsite transport of nutrients and pesticides in runoff. Easton et al (4) reported that the establishment of turfgrass on bare soil increased soil infiltration by more than 65% over a two-year period. As shoot density increased, infiltration rate increased and runoff decreased. Nonetheless, turfgrass sites can contribute to nutrient and pesticide losses to surface water in concentrations greater than recommended. It is the turfgrass manager's responsibility as an environmental steward to practice management techniques that limit runoff transport of potentially dangerous nutrients and pesticides.

Role of Turfgrass Cultural Practices as Deterrents to Runoff

Fertilizer Application

Applying more fertilizer than needed for adequate turfgrass growth and development increases the potential for nutrient runoff. Soil testing measures the nutrient availability in soils, therefore, it is useful for preventing nutrient deficiencies and over-fertilization. Beard (27) states that soil testing is helpful for assessing P, but is not effective for determining N status in turf. Nitrogen tends to give an immediate, favorable response when correctly applied to turfgrass. This favorable greening effect may encourage turf managers to over-apply fertilizer nutrients. Consequently, turfgrass managers must be particularly careful not to apply more N than necessary to meet expectations.

Kunimatsu et al (28) studied the loading rates of nutrients discharged from a golf course in Japan. Samples were taken from a stream that ran from a forested basin through the golf course. They found that an increase in nutrient discharge from the stream was due primarily to the N, P and K applied to the golf course as fertilizer. They also found that the total N and P discharged from the golf course in runoff were 2.5 and 23 times greater, respectively, than the amount discharged from the forested basin. Mallin et al (29) found significant increases in nitrate, ammonium, total P, and DRP in the outflow of a golf course pond compared with the inflow. However, appropriate modern golf course designs usually include containment features that restrict off-site water transport and facilities for reuse of onsite water. A good design can be very effective. Ryals et al (30) investigated three golf courses in North Carolina for nutrient and pesticide contamination. They found chlorothalonil (tetrachloroisophthalonitrile), chlorpyrifos, and 2,4-D residues in on-course water features, but all of the pesticide detections were below environmental hazard levels. Although pesticides were found at each course, none of the outflows contained pesticides at detectable levels. The highest onsite nitrate concentration was 2 mg L^{-1}, with all other samples occurring below 0.7 mg L^{-1}. Onsite concentrations of phosphorus did not exceed 10 μg L^{-1} and averaged 0.09 μg L^{-1} at the course with the highest concentration. Kohler et al (*31*), in a 4-year study at Purdue University's Kampen Golf Course, demonstrated that the construction of a golf course wetland could store and filter runoff from the golf course and surrounding areas, improving the quality of the water entering the course from the surrounding watershed compared with the water quality exiting the course.

Petrovic (*32*) discussed the various fates of N fertilizers applied to turfgrass. Nitrogen is the most commonly used fertilizer for turfgrass maintenance. Nitrogen can be taken up by the plant, lost to the atmosphere, stored in the soil, leached through the soil, or lost through surface runoff (*11*). There are many factors that influence the amount of N that a turfgrass plant can use. These include temperature, moisture, amount of available N, the source and rate of N applied, mowing height and the genetic makeup of the turfgrass plant. When more N is applied than the plant can take up, there is a potential for N loss to the environment. A sound fertilizer program is required of a successful turfgrass manager. However, the manager must be flexible enough to alter the program according to both environmental conditions and plant need, in order to be a good environmental steward as well as a good agronomist.

According to Brown et al (*13*), the potential for N loss to surface or ground water was greatest during periods of low plant use and heavy rainfall. The biological activity of plants is affected by temperature, therefore by season (*33*). Biological activity also differs among species and sometimes among cultivars (*34*). It is up to the turfgrass manager or homeowner to determine the proper fertilizer rates and application timing that achieves the aesthetic and functional value expected of a turfgrass stand without exceeding the amount of fertilizer that the plant can use. The ability of grasses to assimilate fertilizers can also be affected by stress. For instance, Sills and Carrow (*35*) found that the ability of perennial ryegrass (*Lolium perenne* L.) to assimilate fertilizer decreased by up to 30% on compacted soil. Other environmental stresses can also limit the ability of turf to assimilate fertilizers. The applicator should be proficient enough in the effects of environmental factors to adjust fertilizer applications according to plant need.

Pesticide Application

Herbicides are the most commonly used pesticides in turf (*36*). Nearly all turfgrass herbicides are labeled for application to healthy turf. Conversely, insecticides and fungicides are most commonly applied during periods of turfgrass stress. Applicators should be careful to follow label directions for these pesticides explicitly. Pesticides should only be applied when environmental or symptomatic conditions warrant application. Unnecessary pesticide applications are an environmental hazard.

Although research suggests that pesticide applications to turfgrass are not a major environmental pollutant (*6*), losses of these chemicals to surface water can occur when the right circumstances are present. The potential for pesticide runoff is greater for some pesticides than others, and applicators should be familiar with those that are most susceptible to runoff transport. High solubility and persistence are two important factors that contribute to the potential runoff loss of a particular pesticide (*19, 37*). These characteristics can be found on the material safety data sheets (MSDS) available from the pesticide manufacturer and required of all professional applicators. Danger to aquatic organisms is another factor that should be considered when choosing a pesticide, as well as its danger to humans and animals. These precautions can be found on the pesticide label. The effectiveness of a pesticide should also be considered. Many pesticides are not particularly effective under certain weather conditions. These characteristics can be found on the pesticide label and information about pesticide efficacy in a particular geographic region is available from most county extension agents and/or university extension turfgrass specialists. Applying a pesticide at a time or in an amount that is not effective is a wasted application, therefore, an environmental hazard. It is illegal to apply a pesticide at a rate greater than the rates listed on the label. In many states, it is also illegal to apply at a lower rate than those listed on the label. It is also illegal to apply pesticide without reading the label. Obviously, proper application is important for optimal pest control and for environmental protection.

Turfgrass Species Selection

Linde et al (*2*) compared surface runoff from creeping bentgrass (*Agrostis stolonifera* L.) and perennial ryegrass plots maintained under golf course fairway conditions. They analyzed water samples for NO_3-N, PO_4-P, and total Kjeldahl-N. They found that creeping bentgrass reduced runoff volume more than perennial ryegrass, but that the concentrations and loading rates of nutrients in runoff generally did not significantly exceed those found in the irrigation water alone. In another study, Linde et al (*38*) compared surface runoff from creeping bentgrass and perennial ryegrass plots maintained as a golf course fairway. They again found that creeping bentgrass reduced runoff volume more than perennial ryegrass. They attributed this to the higher tiller density of the creeping bentgrass when compared with perennial ryegrass. They reported that the higher tiller density of the creeping bentgrass increased the tortuosity and hydraulic resistance to flow when compared with the perennial ryegrass. In a similar study, Linde et al (*39*) again assessed surface runoff between creeping bentgrass and perennial ryegrass plots and concluded that the selection of a high tiller density, thatch-forming turfgrass such as creeping bentgrass would reduce surface runoff compared to a low tiller density, low thatch-forming turfgrass such as perennial ryegrass. These results were supported by researchers who tested phosphorus and sediment runoff from these two species after aerification (*26*). Steinke et al (*40*) found that Kentucky bluegrass (*Poa pratensis* L.) significantly reduced runoff compared with a relatively open prairie grass mixture, further suggesting that a dense turf provides more runoff resistance than a relatively open canopy. Based on these results, dense growing, thatch- and sod- forming grasses such as creeping bentgrass, Kentucky bluegrass, common (*Cynodon dactylon* L.) and hybrid bermudagrass, and zoysiagrass (*Zoysia japonica* Steud.; *Z. matrella* [L.]; *Z. tenuifolia* Willd.) would be the best choices for cover on slopes that are likely to produce runoff.

Application Timing

Application timing is an important consideration in executing an application management program designed to reduce the transport of nutrients and pesticides in runoff. An indepth discussion of nutrient and pesticide fate following application can be found in reviews by Petrovic (*41*) and Walker and Branham (*5*). Although chemical fate is more likely to be affected by leaching on high sand content surfaces such as many golf course greens and athletic fields, most offsite movement of nutrients and pesticides from turfgrass occurs in runoff. Shuman (*42*) made fertilizer applications followed by precipitation at 4, 24, 48, and 72 hours. The study found that runoff during the first precipitation event following application contained considerably more N and P than the subsequent events. The same pattern is true of pesticide loss (*17, 18*). Consequently, application timing is critical to avoid potential nutrient and pesticide losses to surface runoff.

Weather conditions and weather forecasts are important to the applicator. Wind speed, dew points, temperature and other factors are important for scheduling applications. Post-application precipitation forecasts are also important. An application followed closely by a major rainfall event can result in nutrient or pesticide losses to surface runoff. Conversely, a minor rainfall event that closely follows a nutrient application, pre-emergent herbicide application, insecticide application and some other pesticide applications, and that does not produce runoff helps to prevent chemical losses in subsequent runoff events. Shuman (*43*) tested nutrient runoff from turf when 0.64 cm of irrigation was applied immediately following fertilization, a practice commonly referred to as "watering in" the fertilizer. In that study, subsequent runoff events caused the loss of up to three times the P and twice the N from turf where the fertilizer was not "watered in" compared with turf where the fertilizer was "watered in". Where irrigation is available or handwatering can be practiced, a small amount of irrigation should be applied to "water in" fertilizers and some pesticides. As a matter of protocol, especially on large areas where "watering in" is not practical, turf managers should avoid nutrient and pesticide applications shortly before major storms are predicted.

Although the concentration of the pollutant in runoff is important to the amount of nutrient or pesticide load that will occur from a runoff event, the volume of runoff that occurs is usually the primary factor that determines the amount of nutrient or pesticide lost during an event (*44*). Wauchope et al. (*24*) applied varying formulations of cyanazine (2-[[4-chloro-6-(ethylamino)-1,3,5-triazin-2-yl)]amino]-2-methylpropanenitrile) and sulfometuron-methyl (methyl-2 [[[[(4,6-dimethyl-2-pyrimidinyl)amino]coarbonyl]amino]sulfonyl]benzoate) and found that fractional losses of the pesticides in runoff were linearly related to the amount of pesticide applied, and that differences in concentration were almost entirely accounted for by differences in application rates. These results support those of an earlier study that investigated the amount of atrazine [(6-chloro-*N*-ethyl-*N*'-isopropyl-[1,3,5]triazine-2,4-diamine) likely to occur in runoff (*45*). Consequently, application rate is the primary factor that affects the chemical concentration in runoff from a turfgrass site. The pesticide formulation, liquid or granule, may also affect concentration with the granule resulting in lower concentrations in runoff. The granule, however, is more likely to persist, increasing the potential for loss to occur over time (*46*). Although this formulation effect is generally accepted, it is not consistent in all studies.

Horst et al (*47*) suggested that pesticides are likely to degrade more rapidly in a turfgrass environment than would typically occur in agronomic cropping systems. The researchers found that four commonly used turfgrass pesticides were 50% dissipated within 7 to 16 days after application, depending on the chemical tested. Gardner and Branham (*48*) studied the degradation of mefenoxam [*N*-(2,6-dimethylphenyl)-*N*-(methoxyacetyl)-*D*-alanine methyl ester] and of propiconazole (1-[[2-(2,4-dichlorophenyl)-4-propyl-1,3-dioxolan-2yl]methyl]-1*H*-1,2,4-triazole) in turfgrass and in bare soil. The half-life of mefenoxam was 5-6 days in turf and 7-8 days in bare soil. The half-life of propiconazole was 12-15 days in turf and 29 days in bare soil. Watschke et al (*49*) tested three common turfgrass chemicals in runoff from turfgrass plots and found that the longest persisting chemical was 80% dissipated within four weeks of application. Chemical residues from the other two pesticides were not detectable after three weeks. As the period between a chemical application and a runoff event increases, the potential for nutrient or pesticide losses in runoff declines. The presence of N and P in the turf/soil system will presumably decrease through plant uptake and other factors as the time between application and runoff increases. The key to application timing is to make the application when a post-application runoff event is unlikely to occur for the longest possible period.

Secondary Cultural Practices

There is a wide selection of fertilizer formulations available for use by turfgrass managers. Research suggests that there is little difference in the amount of nutrients lost to runoff from liquid versus granular urea fertilizer (*21*). However, there is reason to believe that slow-release sources of N, those having poor solubility or extended release mechanisms, have a lower leaching potential than highly soluble, quick-release sources (*50*). It seems reasonable that a controlled-release N or P fertilizer source would be less likely to contribute high concentrations of nutrient runoff if its application was followed closely by a major rainfall event. Gaudreau et al (*12*), for instance, found that DRP concentration in runoff three days after applications was five times greater for fertilizer P than for manure P. Brown et al (*51*) demonstrated that N loss from highly soluble fertilizers such as NH_4NO_3 and $(NH_4)_2SO_4$ was greater than that for slow release sources, ureaformaldehyde and Milorganite during leaching and runoff from golf course greens. Fertilizer programs vary widely among turf managers based primarily on turf quality expectations and personal preference. Some controlled-release fertilizers are not very effective during cool weather but work very well during most of the growing season. Use of controlled-release fertilizers whenever possible could reduce the potential for nutrient losses in surface runoff.

Not all management techniques that are conducive to reducing runoff are the result of scientific research. A survey of landscape management practices in Georgia reported that three of four homeowners managed their own landscaping and fully controlled fertilizer and pesticide applications (*52*). Although professional turfgrass managers receive training in application management, many homeowners do not. A survey of four communities in North Carolina revealed that about 60% of the homeowners applied pesticides to their lawns, and that 78% of those calibrated their equipment prior to application (*36*). In contrast, homeowners in these four communities plus one other could usually report the amount of fertilizer applied but rarely could report the rate applied. Only 20% of the homeowners based their fertilization activities on soil tests and even the lawncare companies employed tended to fertilize during the wrong season for warm-season grasses. Based on these results, more consumer education may be needed to help reduce improper application of fertilizers.

Both professional managers and homeowners can reduce the potential for nutrient and pesticide runoff by practicing common sense management. For instance, the application of fertilizer or pesticide to frozen turf and soil is not recommended. Frozen soil is nearly impervious to water and highly likely to produce runoff during precipitation (40). During a study in Wisconsin over a 23-month period, approximately 75% of the runoff recorded from turf occurred when the soil was frozen (53). When soil and turf are frozen, nutrients and pesticides are not likely to bond to those surfaces and easily wash away in runoff.

Nutrient or pesticide applications to saturated soil also increase the potential for environmental contamination. Nutrient or pesticide applications to saturated or nearly saturated soil are likely to be lost to surface runoff if rainfall occurs. Cole et al (1) reported increases in nutrient runoff from 2% of applied to 10% of applied and pesticide losses of 3% of applied that increased to 15% of applied during rainfall simulations on a dry soil site compared with the same site 7 days after 165 mm of rainfall. Shuman (42) found a significant linear relationship between antecedent soil moisture and runoff volume during ten events of 50 mm of simulated rainfall.

Nutrients and pesticides inadvertently applied to surfaces that have relatively low porosity, such as concrete, brick or plastic sheeting, are also likely to wash away in runoff. Stier et al. (53) found that runoff from turfgrass slopes accounted for only seven to nine percent of the runoff collected from concrete slopes. Based on those results, concrete may produce up to 14 times more runoff than turf over any given season.

Aerification

Runoff occurs when the precipitation rate exceeds the infiltration rate of the turf/soil system. Since the infiltration rate of a compacted turf/soil system is lower than the infiltration rate of a similar system that is less compacted, runoff should occur more rapidly and at a higher volume from the most highly compacted site. Core aerification helps to relieve compaction, and should have a positive effect on reducing runoff. However, research to date has not supported that contention. Cole et al (1) and Baird et al (54) reported that core aerification of grass buffers did not affect runoff volume or loss of specific pesticides or nutrients from bermudagrass maintained similar to golf course fairways. Moss et al (55) found that core aerification did not affect runoff volume or nutrient losses from bermudagrass turf and Franklin et al (56) reported that aerification of grasslands managed as hay did not significantly affect the quality and volume of runoff from those sites. Wiecko et al (57) tested five different aerification methods on hybrid bermudagrass grown on a compacted sandy loam soil. Bulk density measurements were made twice following aerification at 0-5 cm and at 5-10 cm depths. Hollow tine aerification significantly reduced soil bulk density at 0-5 cm once in two trials a year apart, and 5-10 cm once in two trials. Solid tine aerification reduced bulk density significantly once at the 0-5 cm depth but did not affect bulk density at the 5-10 cm depth in either year. None of the other aerification methods affected bulk density. However, all but deep drill aerification improved the ease of mechanical penetration into the upper 10 cm of soil with hollow tine aerification outperforming all others. For that reason, and for its agronomic benefits, soil aerification should probably be part of a complete turfgrass management and runoff management program. However, aerification alone does not appear to have a substantial effect on runoff losses.

Attenuation of Runoff Transport with Grass Buffer Strips

The United States Department of Agriculture Natural Resources Conservation Service (USDA-NRCS) defines conservation buffers as small areas or strips of land in permanent vegetation designed to intercept pollutants and manage other environmental concerns (58). Buffers include riparian areas, grassed waterways, shelterbelts, windbreaks, contour grass strips, cross-wind trap strips, field borders and other vegetative barriers. Properly established vegetative buffers may help to reduce or eliminate the movement of sediments (59, 60, 61), pesticides (62, 63, 64) and nutrients (65, 66, 67). Buffers may enhance fish and wildlife habitat, improve or prevent further degradation of water quality, improve soil quality, reduce flooding, conserve energy and conserve biodiversity (58). Grass vegetation is recommended for vegetative buffers or filter strips (68). Buffer performance has been widely studied in agricultural settings and has been effective for reducing pesticide and nutrient losses from agricultural lands (15, 69). However, limited research is found in the area of turfgrass management and buffer performance.

Cole et al (1) studied runoff from bermudagrass plots simulating golf course fairways that were bordered by higher mowed bermudagrass buffers of differing width (0, 1.2, 2.4, and 4.9 m). Average land slopes at the research site were 6%. Urea and super triplephosphate fertilizers, as well as selected herbicides, were applied at normal rates. They found that grass buffers were effective for reducing nutrient runoff when compared to plots without buffers. Their findings indicated that the grass buffers were effective for reducing nutrient and pesticide runoff from golf course fairways, but increasing the buffer width did not affect runoff significantly. Baird et al (54), using the same methods at the same site, determined that a 7.6-cm high bermudagrass buffer bordering a 1.3-cm bermudagrass stand was more effective for reducing runoff than a shorter height buffer of 3.8 cm. During these studies, when the soil was not saturated, less than 2% of applied nutrients were detected in the surface runoff. Nominal losses occurred when 79 mm of simulated rainfall was applied 24 hours following fertilization. The concentrations of nutrients in the runoff, however, were great enough to enhance eutrophication in surface water if not diluted.

In a similar study, Moss et al (44) determined that the effects of grass buffers were enhanced by mowing them at increasingly higher heights. During that study, irrigation was applied to plots bordered by 5.5 m bermudagrass buffers mowed at a single height of 51 mm and 5.5 m bermudagrass buffers divided into 1.8 m segments mowed at increasingly higher heights of 25, 38, and 51 mm. The multiple-height buffers delayed the time from precipitation to runoff by 4 minutes during irrigation, and by 2 minutes during natural rainfall. The volume of runoff that occurred was reduced by 16% during irrigation and 19% during natural rainfall, and the amount of applied N and P fertilizer that was lost in runoff was reduced by an average 16% during irrigation and 14% during natural rainfall using the multiple height buffer compared with the single height buffer. The researchers reported that the interface between higher mown turf at the base of a slope of lower mown turf formed a barrier that slowed runoff increasing soil infiltration time. By mowing the buffers at three different heights, the researchers were, in effect, applying three separate barriers to runoff.

Grasses and vegetative buffers composed of grasses may also work as filters for some pesticides. Krutz et al (70) reported that buffalo grass [*Buchloe dactyloides* (Nutt.) Engelm.] retained atrazine, reducing the loss of the pesticide during simulated rainfall runoff. A dense turfgrass system includes an organic soil layer called thatch that is unique to grasses and has a filtering effect on some pesticides (16). Thatch may also reduce N loss from turf (71). Roy et al (72) found that the degradation rate of dicamba herbicide was 5.9 to 8.4 times greater in thatch than in soil, with a calculated half-life as low as 5.5 days. Thatch retains many pesticides. Horst et al (47) made applications of pendimethalin [N-(1-ethylpropyl)-3,4-dimethyl-2,6-dinitrobenzenamine], chlorpyrifos [O,O-diethyl-O-(3,5,6-trichloro-2-pyridinyl)phosphorothioate], isazaphos {O-[5-chloro-1-(1-methylethyl)-H-1,2,4-triazol-3-yl]O,O-diethyl phosphorothioate}, and metalaxyl [N-(2,6-dimethylphenyl)-N-(methoxyacetyl) alanine methyl ester] to turfgrass at two locations over two years and measured pesticide retention in verdure, thatch, and soil during most of two growing seasons. They found that the most hydrophobic pesticides, pendimethalin and chlorpyrifos, were highly retained in the thatch but the more hydrophyllic pesticides, isazofos and especially metalaxyl, were more mobile. No more than 3% of the pendimethalin and 5% of the chlorpyrifos initially present after application moved through the thatch to underlying soil. In contrast, more than 28% of the metalaxyl residue recovered was found in the soil profile. Thatch can be a factor in pesticide and nutrient losses from turf. The combination of pesticide retention and flow rate reduction enables grass buffers to reduce runoff losses from crop fields or turfgrass.

Recommendations

Based on research to date, turf inhibits runoff better than any other surface. Even some manmade materials specifically designed to reduce erosion do not inhibit runoff as well as turf. However, based on previous studies, runoff from turf does occur and can potentially contain unacceptable levels of nutrients and pesticides. Turfgrass managers must be aware of and execute management practices that help to reduce this potential environmental contamination. Applications of fertilizers or pesticides may be important for managing a healthy dense turf that resists runoff flow. However, improper application practices or application rates that exceed those warranted can be dangerous to the environment. Applications of fertilizers and pesticides should not be made unless soil and/or tissue tests, integrated pest management programs, and/or turfgrass growing conditions warrant their use. Turfgrass managers must be adequately educated in the use of these products and must be flexible enough to adjust application timing and rates to environmental conditions that affect turfgrass growth and development. Maintaining dense turf inhibits runoff but over-fertilization or unnecessary pesticide applications designed to improve turf density beyond reasonable expectations should be avoided. Applications to saturated soil, low infiltration rate soils, soils with a shallow water table that are often wet, frozen soil, or relatively impermeable materials are likely to cause nutrient and pesticide losses to runoff during subsequent rainfall events. Quick-release nitrogen fertilizer sources are more likely to produce high N concentrations in runoff than slow-release sources if runoff occurs shortly after application. There is some evidence that granular insecticides and herbicides may be less likely to produce chemical runoff in high concentrations than liquid-applied products. "Watering in" fertilizers or pesticides when recommended with a small amount of irrigation or rainfall that does not produce runoff is effective for reducing the amount of product lost to a runoff event that occurs shortly after application.

Aggressive thatch producing turfgrass species that spread by rhizomes and/or stolons such as bermudagrass, zoysiagrass, Kentucky bluegrass, and creeping bentgrass are more likely to inhibit runoff than bunch-type grasses. Maintaining an acceptable thatch layer that does not seriously affect turfgrass performance can help to mitigate the potential for offsite contamination. Although scientific studies have not adequately determined that core aerification reduces runoff volume or nutrient and pesticide loss to runoff, aerification should be considered in an overall turfgrass management plan.

Research has demonstrated that grass vegetation buffers, especially those mowed at multiple mowing heights, are effective for reducing runoff losses.

The rate of application, the solubility of the product applied, and the length of time between product application and a runoff-producing precipitation event are the primary factors that determine the chemical concentration of the product in runoff, when runoff occurs. The initial runoff event following application will contain substantially higher concentrations than subsequent events and is consequently, of greatest concern. Therefore, application timing is critical. Weather can be unpredictable, but the applicator should endeavor to apply products at a time when a major storm event is not likely to occur within a few days after application.

Once runoff begins, the concentration of product in the runoff can no longer be affected and the amount of product lost is determined by the volume of runoff that occurs. The turfgrass management practices employed prior to a runoff event can affect runoff volume. Selecting the best turfgrass species for the site, using cultural practices that encourage dense turf and an adequate thatch layer, core aerification, and providing grass buffer strips where possible are examples of management practices that help to reduce runoff volume.

A healthy, dense turfgrass is a runoff deterrent. However, the effectiveness of that deterrent can be enhanced with practical application management strategies and cultural management programs that specifically address potential nutrient and pesticide transport in runoff.

References

1. Cole, J.T.; Baird, J.H.; Basta, N.T.; Hunke, R.L. ; Storm, D.E.; Johnson, G.V.; Payton, M.E; Smolen, M.D.; Martin, D.L.; Cole, J.C. *J. Environ. Qual.,* **1997**, *26*, 1589-1598.
2. Linde, D.T.; Watschke, T.L.; Borger, J.A. In *Science and Golf II: Proceedings of the World Scientific Congress of Golf;* Cochran, A.J.; Farrally, M.R., Ed.; E and FN Spon, London, 1995, pp 489-496.
3. Easton, Z.M.; Petrovic, A.M. *J. Environ. Qual.* **2004**, *33*, 645-655.
4. Easton, Z.M.; Petrovic, A.M.; Lisk, D.J.; Larsson-Kovach, I.M.; *Int. Turfgrass Soc. Res. J.* **2005**, *10*, 121-129.
5. Walker, W.J.; Branham, B. In *Golf Course Management and Construction: Environmental Issues;* Balogh, J.C.; Walker, W.J. Ed.; Lewis Publ., Chelsea, MI, 1992, pp 105-219.
6. Cohen, S.; Svrjcek, A.; Durborow, T.; Barnes, N.L. *J. Environ. Qual.* **1999**, *28*, 798-809.
7. Daniel, T.C.; Sharpley, A.N.; Lemunyon, J.L. *J. Environ. Qual.* **1998**, *27*, 251-257.
8. Rosen, C.J.; Horgan, B.P. *Int. Turfgrass Soc. Res. J.* **2005**, *10*, 130-135.
9. NOAA/EPA. *Strategic Assessment of Near Coastal Water.* NOAA, Washington, D.C., 1988.

10. USEPA. *Quality Criteria for Water*. U.S. Gov. Print. Office, Washington, DC, 1976.
11. Brady, N.C. *The Nature and Properties of Soils*, 10th Ed.; Macmillan: New York, NY, 1990.
12. Gaudreau, J.E.; Vietor, D.M.; White, R.H.; Provin, T.L.; Munster, C.L. *J. Environ. Qual.*. **2002**, *31*, 1316-1322.
13. Brown, K.W.; Duble, R.L.; Thomas, J.C. *Agron J.* **1977**, *69*, 667-671.
14. Burton Jr., G.A.; Pitt, R.E. *Stormwater Effects Handbook*. CRC Press, Boca Raton, FL, 2002.
15. Sharpley, A.; Foy, B.; Withers, P. *J. Environ. Qual.* **2000**, *29*, 1-9.
16. Raturi, S.; Carroll, M.J.; Hill, R.L. *J. Environ. Qual.* **2003**, *32*, 215-223.
17. Smith, A.E.; Bridges, D. C. *Crop Sci.* **1996**, *36*, 1439-1445.
18. Ma, Q.L.; Smith, A.E.; Hook, J.E.; Smith, R.E.; Bridges D.C. *J. Environ. Qual.* **1999**, *28*, 1463-1473.
19. Smith, A.E. *Int. Turfgrass Soc. Res. J.* **1997**, *8*, 197-204.
20. Krenitsky, E.C.; Carroll, M.J.; Hill, R.L.; Krouse, J.M. *Crop Sci.* **1998**, *38*, 1042-1046.
21. Gross, C.M.; Angle, J. S.; Welterlen, M. S. *J. Environ. Qual.* **1990**, *19*, 663-668.
22. Gross, C.M.; Angle, J.S.; Hill, R.L.; Welterlen, M.S. *J. Environ. Qual.* **1991**, *20*, 604-607.
23. Linde, D.T., Watschke, T.L.; Jarrett, A.R. *J. Environ. Qual.* **1997**, *26*, 1248-1254.
24. Wauchope, R.D.; Williams, R.G.; Luz, R.M. *J. Environ. Qual.* **1990**, *19*, 119-125.
25. Harrison, S.A.; Watschke, T.L.; Mumma, R.O.; Jarrett, A.R.; Hamilton, Jr., G.W. In *Pesticides in Urban Environments: Fate and Significance*; Racke, K. D.; Leslie, A.R., Ed. American Chemical Society, Washington, DC., 1993, pp 191-207.
26. Kauffman III, G.L.; Watschke, T.L. *Agron. J.* **2007**, *99*, 141-147.
27. Beard, J.B. *Turf Management for Golf Courses*, 2nd ed; Ann Arbor Press, Chelsea, MI, 2002.
28. Kunimatsu, T.; Sudo, M.; Kawachi, T. *Water Sci. Technol.* **1999**, *39*, 99-107.
29. Mallin, M.A.; Ensign, S.H.; Wheeler, T.L.; Mayes, D.B. *J. Environ. Qual.* **2002**, *31*, 654-660.
30. Ryals, S.C.; Genter, M.B.; Leidy, R.B. *Environ. Toxicology Chem.* **1998**, *17*, 1934-1942.
31. Kohler, E.A.; Poole, V.L.; Reicher, Z.J.; Turco, R.F. *Ecological Engineering*. **2004**, *23*, 285-298.
32. Petrovic, A.M. *J. Environ. Qual.* **1990**, *19*, 1-14.

33. Leopold, A.C.; Kriedman, P.E. *Plant Growth and Development*, 2nd Ed.; McGraw-Hill: New York, NY, 1975.
34. Treshow, M. *Environment and Plant Response*; McGraw-Hill: New York, NY, 1970.
35. Sills, M.J.; Carrow, R.N. *Agron J.* **1983**, *75*, 488-492.
36. Osmond, D.L.; Hardy, D.H. *J. Environ. Qual.* **2004**, *33*, 565-575.
37. Haith, D.A.; Rossi, F.S. *J. Environ. Qual.*. **2003**, *32*, 447-455.
38. Linde, D.T.; Watschke, T.L.; Jarrett, A.R.; Borger, J.A. *Agron. J.* **1995**, *87*, 176-182.
39 Linde, D.T.; Watschke, T.L.; Jarrett, A.R. *J. Turfgrass Manag.* **1998**, *2*, 11-34.
40. Steinke, K.; Stier, J.C.; Kussow, W.R.; Thompson, A. *J. Environ. Qual.*. **2007**, *36*, 426-439.
41. Petrovic, A.M.; Easton, Z.M. *Int. Turfgrass Soc. Res. J.* **2005**, *10*, 55-69.
42. Shuman, L.M. *J. Environ. Qual.* **2002**, *31*, 1710-1715.
44. Moss, J.Q.; Bell, G.E.; Kizer, M.A.; Payton, M.E.; Zhang, H.; Martin, D.L. *Crop Sci.* **2005**, *46*, 72-80.
45. Shuman, L.M. *Comm. Soil Sci. Plant Anal.* **2004**, *35*, 9-24.
45. Wauchope, D. *J. Environ. Qual.* **1987**, *16*, 212-216.
46. Song, H.; Smith A.E. *J. Environ. Qual.* **1997**, *26*, 379-386.
47. Horst, G.L.; Shea, P.J.; Christians, N.; Miller, D.R.; Suefer-Powell, C.; Starrett, S.K. *Crop Sci.* **1996**, *36*, 362-370.
48. Gardner, D.S.; Branham, B.E. *J. Environ. Qual.*. **2001**, *30*, 1612-1618.
49. Watschke, T.L.; Mumma, R.O.; Linde, D.T.; Borger, J.A.; Harrison, S.A. In *Fate and Management of Turfgrass Chemicals;* Clark, J.M.; Kenna, M.P., Ed. American Chemical Society, Washington, DC, 2000, pp 94-105.
50. Quiroga-Garza, H.M., Picchioni, G.A.; Remenga, M.D. *J. Environ. Qual.* **2001**, *30*, 440-448.
51. Brown, K.W.; Thomas, J.C.; Duble, R.L. *Agron. J.* **1982**. 74, 947-950.
52. Varlamoff, S.; Florkowski, W.J.; Jordan, J.L.; Latimer, J.; Braman, K. *HortTechnol.* **2001**, *11*, 326-331.
53. Stier, J.C.; Walston, A.; Williamson R.C. *Int. Turfgrass Soc. Res. J.* **2005**, *10*, 136-143.
54. Baird, J.H.; Basta, N.T.; Hunke, R.L.; Johnson, G.V.; Payton, M.E.; Storm, D.E.; Wilson, C.A.; Smolen, M.D.; Martin, D.L.; Cole, J.T. In *Fate and Management of Turfgrass Chemicals*; Clark, J.M.; Kenna, M.P., Ed.;. American Chemical Society, Washington, DC, 2000, pp 268-293.
55. Moss, J.Q; Bell, G.E.; Martin, D.L.; Payton, M.E. *J. Applied Turfgrass Science.* **2007**, Online, doi:10.1094/ATS-2007-0125-02-RS.
56. Franklin, D.H.; Cabrera, M.L.; Calvert, V.H. *Soil Sci. Soc. Am. J.* **2006**, *70*, 84-89.

57. Wiecko, G.; Carrow, R.N.; Karnok, K.J. *Int. Turfgrass Soc. Res. J.* **1993**, 451-457.
58. USDA-NRCS. *Conservation Buffers to Reduce Pesticide Losses.* 2000. Online,ftp://ftp.wcc.nrcs.usda.gov/downloads/pestmgt/newconbuf.pdf (verified 13 Jul 2007).
59. Barfield, B.J.; Tollner, E.W.; Hayes, J.C. *Trans. ASAE.* **1979**, *22*, 540-548.
60. Hayes, J.C.; Barfield, B.J.; Barnhisel, R.I. *Trans. ASAE.* **1979**, *22*, 1063-1067.
61. Robinson, C. A.; Ghaffarzadeh, M.; Cruse, R.M.J. *Soil Water Conserv.* **1996**, *50*, 227-230.
62. Misra, A.K.; Baker, J.L.; Mickelson, S.K.; Shang, H. *Trans. ASAE.* **1996**, *39*, 2105-2111.
63. Patty, L.; Real, B.; Gril, J.J. *Pestic. Sci.* **1997**, *49*, 243-251.
64. Mersie, W.; Seybold, C.A.; McNamee, C.; Huang, J. *J. Environ. Qual.* **1999**, *28*, 816-821.
65. Lee, K-H.; Isenhart, T.M.; Schultz, R.C.; Mickelson, S.K. *Agroforest. Sys.* **1999**, *44*, 121-132.
66. Magette, W.L.; Brinsfield, R.B; Palmer, R.E.; Woods, J.D. *Trans. ASAE.* **1989**, *32*, 663-667.
67. Mendez, A.; Dillaha, T.A.; Mostaghimi, S. *J. Am. Water Res. Assoc.* **1999**, *35*, 867-875.
68. USDA-NRCS. *National Handbook of Conservation Practices.* Item no. 0120-A, USDA-NRCS, Washington, DC, 1997.
69. Baker, J.L., Mickelson, S.K.; Arora, K.; Misra, A.K. In *Agrochemical Movement: Perspective and Scale.* Steinheimer, T.R., Ed; American Chemical Society, Washington, DC, 2000, pp 272-287.
70. Krutz, L.J.; Senseman, S.A.; Dozier, M.C.; Hoffman, D.W.; Tierney, D.P. *J. Environ. Qual.* **2003**, 2319-2324.
71. Engelsjord, M.E.; Branham, B.E.; Horgan, B.P. *Crop Sci.* **2004**, *44*, 1341-1347.
72. Roy, J.W.; Hall, J.C.; Parkin, G.W.; Wagner-Riddle, C.; Clegg, B.S. *J. Environ. Qual.* **2001**, *30*, 1360-1370.

Chapter 9

Sediment and Nutrient Losses from Prairie and Turfgrass Buffer Strips during Establishment

K. Steinke[1], J. C. Stier[2], W. R. Kussow[3], and A. Thompson[4]

[1]Department of Soil and Crop Sciences, Texas A&M University, College Station, TX 77843
Departments of [2]Horticulture, [3]Soil Science, and [4]Biological Systems Engineering, University of Wisconsin at Madison, Madison, WI 53706

The increase in impervious surface areas associated with urban development has created a more efficient water conveyance system leading to heightened surface water runoff flows and volumes. There is public concern about the continued degradation òf bodies of water through increased sediment and nutrient loading, and the effects that urban vegetation may have on these receiving waters. We conducted a study focusing on the qualitative and quantitative characteristics of surface water runoff from urban landscapes. Turfgrass and prairie landscapes were compared for their use as buffer strips in reducing nutrient- and sediment-rich runoff from urban areas. Three impervious to pervious surface ratios (1:1, 1:2, 1:4) were studied to obtain information for improved design of urban landscapes to control runoff. Runoff events from natural precipitation were monitored continuously for 24 months following seeding of vegetated buffer strips downslope from pavement. All water samples were analyzed for phosphorus (P) as total phosphorus, bioavailable phosphorus, soluble phosphorus, organic soluble phosphorus, and for sediment concentration. The majority (>75%) of runoff, nutrient, and sediment loading occurred during frozen soil conditions. A 2:1 or 4:1 buffer:pavement ratio reduced runoff and nutrient loading only during nonfrozen soil conditions. Prairie vegetation had 46% and 119% more total runoff volume than turfgrass vegetation during the first two years of establishment, respectively (p = 0.08). Sediment losses from prairie were two to five times greater than from turf (p < 0.05) during non-frozen soil conditions, and decreased in both systems from year one to year two. Total phosphorus loads averaged 0.18 kg ha^{-1} and 0.08 kg ha^{-1} from native prairie, compared to 0.07 kg ha^{-1} and 0.02 kg ha^{-1} from turfgrass buffers during nonfrozen soil conditions during the first and second years of establishment, respectively (p < 0.05). Total annual P and sediment losses from prairie and turf were similar, however, because runoff during frozen soil conditions was significantly greater than runoff from nonfrozen soil. Although not an intended focus of the study, the high proportion of forbs (broadleaf prairie herbs) relative to prairie grasses in the commercial seed mixtures recommended for urban plantings may not truly represent prairies which existed prior to European colonization. Since only a small portion of runoff occurred from either of the vegetated surface types during the growing season, surface water quality in temperate regions may benefit best from urban stormwater management practices aimed at controlling runoff during the winter months.

The continued rise in urban development has transformed once agriculturally-dominated landscapes into a combination of impervious surfaces and multiple vegetation types. These transformations have lead to increased runoff volumes, nutrient loads, and flooding frequencies and magnitudes (*1, 2, 3*). Urban landscapes continue to be identified as a source of non-point source (NPS) pollution, but few data exist to support recommended best management practices within these locations. As a result, urban surface water quality has continued to decline, and the identification of the source(s) and mechanism(s) behind this method of water eutrophication remains difficult.

The current research project was developed as a result of a proposed state regulation targeting urban runoff and NPS pollution. In 2001, the Wisconsin Department of Natural Resources (WDNR) revised a year 2000 proposal concerning the reduction of runoff pollution (*4*). The proposed legislation included measures to control runoff from urban areas by requiring the use of vegetative buffer strips within these areas to reduce runoff volume and prevent the discharge of sediment and nutrients from these sites into surface waters. The proposal specifically stated that "native vegetation is preferred and short-rooted vegetation such as Kentucky bluegrass is least desirable" (*4*). The legislation was written based on authors' perceptions, as no scientifically derived data existed directly comparing runoff between prairie and turfgrass environments within an urban setting.

Phosphorus is the key nutrient involved in the eutrophication of freshwater lakes and streams (*5*), and is used extensively in turfgrass fertilizers. This had led counties and municipalities to pass restrictions on the use of P-containing turfgrass fertilizers and suggest the use of native plantings in place of turfgrass due to lower maintenance requirements. A large percentage of urban turfgrass areas consist of mineral-based topsoils, often brought in from agricultural lands that have elevated soil P levels and can lead to increased risk for P pollution. While turfgrass requires fertilization and irrigation for best quality, research has shown little nutrient and sediment runoff occurs from properly managed turfgrass (*6, 7*). In fact, ground cover density may be more important than other factors for controlling runoff from surfaces (*8, 9*). Prairie ecosystems are known for their sustainable characteristics from maintaining a balance in nutrient gains and losses over time. Despite being classified as native, and lacking the need for anthropic fertility inputs, these prairie environments still undergo nutrient biogeochemical cycling and can serve as potential sources of nutrient pollution (*10*).

The objective of this study was to evaluate native prairie and turfgrass buffer strip performance by quantifying sediment and P loading from these areas. In addition, the ratio of impervious:pervious surfaces for mitigating urban stormwater runoff and natural contaminants was evaluated.

Materials and Methods

Establishment

Surface water runoff was monitored from field plots developed during autumn 2002, measuring 9.75 m in length by 2.44 m in width. Plots were arranged on a 5.5% slope and outfitted with runoff collectors designed according to Brakensiek et al. (*11*). Runoff collectors were constructed from galvanized steel, measured 2.44 m in width, attached to steel cutoff plates, and were installed at the downslope base of individual runoff plots according to Stier et al (*12*).

Experimental plots were arranged in a randomized complete block design with three replications of a 3 x 2 factorial treatment arrangement. Treatments were assigned to the plots as follows: 1) concrete + turfgrass buffer strip, 1:1 ratio; 2) concrete + prairie buffer strip, 1:1 ratio; 3) concrete + turfgrass buffer strip, 1:2 ratio; 4) concrete + prairie buffer strip 1:2 ratio; 5) concrete + turfgrass buffer strip, 1:4 ratio; 6) concrete + prairie buffer strip, 1:4 ratio. Plots were dormant-seeded during 4 November 2002 and covered with Futerra® wood-fiber mulch blankets (Profile Products LLC, Buffalo Grove, IL). Turfgrass vegetation was a blend of equal percentages of 'Odyssey', 'Arcadia', 'Cynthia', 'Cannon', and 'Showcase' Kentucky bluegrasses (*Poa pratensis* L.). A locally recommended (University of Wisconsin-Madison), readily available prairie mixture common to Upper Midwest prairie restorations was chosen for this study. Prairie vegetation consisted of a native, shortgrass, medium soil mixture consisting of 16 forbs, two legumes, and three warm season (C-4) grasses (Prairie Nursery Inc., Westfield, WI, Stock #50002). Runoff collection occurred after each snowmelt or rainfall event with runoff volumes being recorded and 250 mL samples frozen for later analysis.

Management

Following germination in April 2003, turfgrass was mulch-mowed at a height of 6.35 cm as needed so that no more than 1/3 of the vegetation was removed at one time. Granular 21-3-12 fertilizer was applied using a drop-spreader at a rate of 48 kg nitrogen ha[-1] three times yearly (May, September, and November). Prairie vegetation was maintained at a height of 8.89 cm through June 1, 2003, then cut monthly at a height of 20.32 cm during 16 June, 15 July, and 16 August 2003 to decrease weed competition (*13*). Vegetation was mowed to within 1.27 cm of ground level and biomass removed on 23 April 2004 to simulate burning, as burning is not allowed within city limits. Vegetation was cut for the final time to a height of 30.48 cm on 1 June 2004, to decrease weed competition. Prairie vegetation was not fertilized at any time. All plots relied exclusively on natural precipitation as no artificial irrigation or rainfall simulations were applied to either vegetation type.

Water Analysis

Total P analysis was performed through a persulfate digestion (*14*). An aliquot of the digested solution was analyzed for solution orthophosphate (*15*). Soluble P was determined by centrifuging runoff water at 2500 rpm for 15 minutes followed by analysis of an aliquot of the supernatant for solution orthophosphate. A second aliquot from the supernatant of the centrifuged runoff sample was heated to 500 °C and analyzed for solution orthophosphate to determine soluble organic P. The solution orthophosphate reading for soluble P was subtracted from the solution orthophosphate reading of the second heated sample to determine the soluble organic P concentration. Bioavailable P was determined through shaking iron oxide-impregnated filter paper strips with runoff samples (*16*). Phosphorus extracted from the filter disks was analyzed for solution orthophosphate. Solution orthophosphate for all methods was analyzed using a Spectronic 21D (Milton Roy, Rochester, NY). Total suspended solids were determined by evaporating runoff samples overnight in a drying oven to determine total suspended solids content. The dried solids were then ashed at 500 °C in a muffle furnace to determine organic solids content.

Results and Discussion

Surface Runoff Water Quantity

Due to the dormant plot seeding during Autumn 2002, Spring 2003 snowmelt occurred over soil covered with wood-fiber erosion control mulch blankets, and runoff volume results were not influenced by vegetation or buffer strip size (data not shown). Six of 15 total runoff events in 2003 occurred during the winter snowmelt period providing approximately 2-4 times more runoff for prairie and turf, respectively, than the nine events during the 2003 growing season. In 2004, vegetation was present during snowmelt conditions and 11 of 15 runoff events occurred during spring snowmelt or rainfall over frozen soil conditions. Total runoff volumes for 2004 did not significantly differ among the buffer strip ratios because over 85% of the annual runoff volume occurred during frozen soil conditions (Table I). Both the 1:2 and 1:4 impervious:pervious surface treatments significantly reduced runoff by at least 35% during nonfrozen soil conditions in 2003, and 60% in 2004 ,compared to the 1:1 surface treatment (P < 0.05). Vegetation type did not influence the amount of runoff from any of the buffer strip ratios during either frozen or nonfrozen soil conditions.

Mean annual runoff volumes were similar for turf and prairie (Table I). Over 80% of annual runoff occurred while soil was frozen. While a significantly greater (P = 0.01) percentage of turf runoff (95%) occurred during frozen conditions than did prairie runoff (83%), the reverse was true during the growing season. On an actual volume basis, turf had approximately 30-50% less nonfrozen soil runoff than prairie in 2003 and 2004, respectively (P = 0.08), but similar amounts during frozen conditions resulted in no annual differences.

Mickelson et al (*17*) found buffer length to be a significant parameter in reducing surface water runoff in agricultural systems to meet conservation compliance plans using forage-dominated buffer strips. However, forage grasses have less vegetative density than turf as higher vegetation heights promote lower plant density (*18*). Previous data addressing total annual runoff volumes from turfgrass and prairie vegetation under similar soil and climatic conditions do not exist and few studies have included data on snowmelt runoff.

Table I. Mean Annual Runoff Volumes 2003, 2004 from Verona, WI

	Non-Frozen Soil (L m^{-2})[a]		Frozen Soil (L m^{-2})	Mean Annual Runoff Volume (L m^{-2})	% Volume from Frozen Soil
	2003	2004	2003-2004	2004	2004
Buffer Size[b]					
1:4	14.50a	3.17a	73.00a	76.17a	95.84b
1:2	11.30a	4.29a	45.24a	49.53a	91.34ab
1:1	22.41b	11.47b	79.05a	90.52a	87.33a
P-value	0.04	0.04	0.43	0.38	0.20
Vegetation[c]					
Prairie	19.07	8.66	58.33	66.99	83.33
Turfgrass	13.07	3.95	73.19	77.14	94.88
P-value	0.08	0.08	0.51	0.67	0.01

[a] Units of L m^{-2} are equivalent to mm of runoff.

[b] Plots measured 2.44 m in width by 9.75 m in length, or 23.78m^2, and consisted of impervious:pervious rations of concrete to prairie or turf grass.

[c] Plots were dormant seeded November 2002 and covered with wood-fiber mulch blankets. No vegetation was present during Winter 2002-2003.

White and Williamson (*19*) demonstrated that the amount of pervious surface has little impact on winter runoff volumes, as they speculated that a thin layer of ice often separates runoff water from the soil surface creating an effectively sealed surface.

The 76.8% (bare, erosion mat-covered soil) and 94.9% of runoff from turfgrass occurring over frozen soil during years one and two of the current study are comparable to the 68.5% to 99.0% recorded in another Wisconsin study (*20*) and 62.0% recorded in New York (*21*). Runoff due to snowmelt or rainfall over frozen prairie soil accounted for 61.2% (bare, erosion mat-covered soil) and 83.3% of total prairie runoff during the first two years of the study. These values are also comparable to the 80.0% of runoff accredited to snowmelt from a native Minnesota prairie (*10*). Our lack of significant treatment differences on frozen soil indicates that the amount of pervious surface is insignificant during the winter months, rendering the impervious:pervious surface ratios irrelevant. For urban landscapes, our data suggest buffer size only influences runoff events occurring during nonfrozen soil conditions. Buffer strip to impervious surface ratio may play a more integral role in reducing runoff volumes in southern climates where soil does not typically freeze.

Total Suspended Solids

Neither buffer strip ratio nor vegetation type significantly affected the annual total suspended solids (TSS) loads during 2003 or 2004 (data not shown). However, significant differences were apparent between vegetation types during both years during nonfrozen soil conditions (Figure 1). Turfgrass lost less TSS than prairie in both years during nonfrozen soil conditions ($P < 0.05$); differences were affected by vegetation type due to establishment rate differences. Native prairie vegetation lost 1.8 and 7.1 times greater TSS quantities than turfgrass vegetation during the first and second years of the study, respectively. The large increase in TSS losses from prairie vegetation relative to turf during year two was not a function of an actual increase in measured losses but rather a function of a dramatic decrease in the TSS losses from turfgrass vegetation due to its more rapid establishment rate. As demonstrated in Figure 1, the measured TSS values from prairie vegetation decreased from years one to two (30.9 vs. 26.3 kg ha^{-1} yr^{-1}). However, TSS loading from turfgrass vegetation decreased from 17.5 to 3.7 kg ha^{-1} yr^{-1} during the first and second years, respectively.

Turfgrass vegetation allowed significantly less organic solids loss during both years of the study as compared to the prairie (Figure 1). Organic TSS decreased in both vegetation systems over time. The percentage of TSS consisting of organics increased for turfgrass vegetation from years one to two (27.8% vs. 30.0%) whereas the percentages decreased for prairie vegetation (39.3% to 21.3%) during this same time period.

158

During both years of the study, mean TSS loading by season was statistically significant (P < 0.10) but probably not biologically significant for 3 of the 8 climatic seasons (Figure 2). The bulk of the TSS erosion occurred during the winter months (December – March) for both vegetation types.

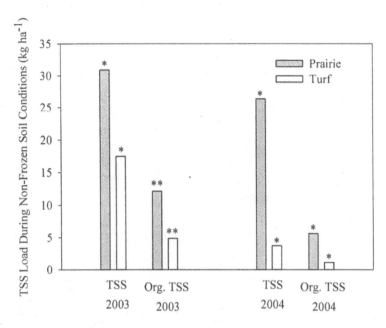

*Figure 1. Comparison of total suspended solids (TSS load during nonfrozen soil conditions between turfgrass and native prairie vegetation 2003-2004, Verona, WI. * and ** indicate significance at the α = 0.05 or 0.01 probability levels, respectively.*

Winter TSS loading accounted for 84.7 and 73.6% of the prairie and 91.6 to 95.3% of the turfgrass TSS loads during years one and two, respectively (Figure 2). This trend strongly parallels what is seen in Table I pertaining to the percentage of runoff occurring over frozen soil and may suggest a relationship.

Few data concerning soil erosion during the establishment year of prairie or turfgrass vegetation are available. White and Williamson (*19*) measured runoff sediment losses from an established relict prairie in South Dakota and found erosion to be too small to adequately measure. The researchers did state that erosion losses from pristine prairie would potentially be similar to those obtained from cultivated land, since prairie fires were not controlled under pristine conditions, and erosion from recently burned land is typically greater

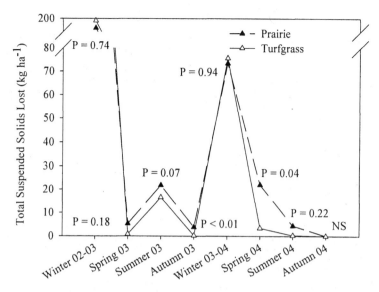

Figure 2. Timing of erosion losses in turfgrass and native prairie vegetation 2003-2004, Verona, WI.

than from unburned land. Despite a wealth of turfgrass runoff data, few studies have measured runoff and sediment loading during plant establishment. Krenitsky et al (*8*) demonstrated turfgrass sod to be an effective erosion control material which lost 60 to 250 times less sediment compared to bare soil. Our 3.7 kg ha^{-1} yr^{-1} TSS loss from turfgrass during the second year of the study (first established year) was less than the 5.5 to 14.4 kg ha^{-1} yr^{-1} observed in Maryland from a well established, fertilized turfgrass stand (*22*). These differences may be due in part to growth habit as the Maryland study had 60% tall fescue (*Festuca arundinacea* Schreb.) in the intial seeding, which has a bunch-type growth habit. The 100% Kentucky bluegrass in the current study had a rhizomatous growth habit which may stabilize soil conditions through the spread of below surface rhizomes and thatch development to prevent excess sediment loss from occurring. Linde et al (*9*) previously reported that the bunch-type growth habit of perennial ryegrass (*Lolium perenne* L.) allowed more runoff than thatch-producing creeping bentgrass (*Agrostis palustris* Huds.)

Surface Water Quality

Approximately 89.8% and 97.4% of the total P loading for the prairie and turfgrass landscapes, respectively, occurred over frozen soil (Table II).

Phosphorus losses (total, soluble, soluble organic and bioavailable P) were significantly influenced by vegetative treatment during the nonfrozen soil periods of 2003 and 2004 (Table II). All types of P losses were usually significantly greater ($P < 0.05$) from prairie vegetation than turf during unfrozen conditions. However, as with runoff volume, the vast majority of P losses occurred during frozen soil conditions resulting in similar annual losses between the two systems (Table II).

Phosphorus loading from nonfrozen soil for all P forms sharply decreased from project years one to two (Table II), demonstrating the effects of plant growth on nutrient runoff losses. These losses occurred on relatively immature turfgrass and prairie vegetation, but the losses need to be documented if an effective nutrient management plan is to be constructed. Total P, consisting of the total amount of soluble and sediment bound P in water, may be the best indicator of lake water quality (5). Mean turfgrass total P loading for 2004 was maintained at 0.66 kg ha^{-1}, which is greater than the 0.006 to 0.033 kg ha^{-1} achieved in a Maryland study (22). These differences between studies may be credited towards differences in precipitation patterns, less mature vegetation, or species composition. The dominance of snowmelt runoff may explain the decreased TSS load, but our total P load was 10 fold greater compared to the Maryland study, where winter runoff was not as large of a percentage of total annual runoff. The large quantities of water running off during the Winter may compensate for the decreased total P concentrations, resulting in higher nutrient loading. Mean prairie total P loading during 2004 was 0.82 kg ha^{-1}. This value was much larger than a Minnesota study (10) which recorded a mean total P load of 0.11 kg ha^{-1} over a 5-year period. The Minnesota study occurred on an established, grass-dominated prairie while the current study took place on a newly-established, forb-dominated prairie. According to Tables I and II, an increase in total P nutrient load collected from both vegetation types during the winter period paralleled the frozen soil runoff volumes, and may indicate an association between Winter total P loadings and Winter runoff volumes.

Soluble P can have an immediate impact on plant growth and is the form of P most available for plant uptake in the soil solution. Soluble P loading from turfgrass in the present study was 0.45 kg ha^{-1} for 2004, much greater than the 0.003 to 0.020 kg ha^{-1} observed in the Maryland study. However, the percentage of soluble P was very similar between the two studies, with soluble P comprising 67.6% of the total P in the current study compared to 57.5% in the Maryland study. Prairie soluble P loading was 0.55 kg ha^{-1}, much greater than the Minnesota study. However, soluble P comprised 67.1% of the total P in the current study, compared to only 27.3% in the Minnesota study. Soluble organic P was a minimal contributor to the soluble and total P nutrient load; nonetheless this fraction of P did represent close to 7.5% and 2.5% of the turfgrass and prairie nutrient loading, respectively (Table II).

Bioavailable P represents that fraction of P encompassing all soluble P and the portion of sediment-attached P that can come into solution. The bioavailable

Table II. Mean Phosphorus Fractionation and Loading 2003-2004 from Verona, WI

	Total P				Soluble P			
	Nonfrozen Soil (kg ha^{-1})		Frozen Soil^b (kg ha^{-1})	Mean Annual Nutrient Load (kg ha^{-1})	Nonfrozen Soil (kg ha^{-1})		Frozen Soil (kg ha^{-1})	Mean Annual Nutrient Load (kg ha^{-1})
	2003	2004	2003-2004	2004	2003	2004	2003-2004	2004
Vegetation[a,b]								
Prairie	0.1756	0.0830	0.7328	0.8158	0.1349	0.0359	0.5116	0.5475
Turfgrass	0.0739	0.0171	0.6478	0.6649	0.0586	0.0090	0.4402	0.4492
P-value	<0.01	0.02	0.68	0.49	<0.01	0.01	0.65	0.54

	Soluble Organic P				Bioavailable P			
	Nonfrozen Soil (kg ha^{-1})		Frozen Soil (kg ha^{-1})	Mean Annual Nutrient Load (kg ha^{-1})	Nonfrozen Soil (kg ha^{-1})		Frozen Soil (kg ha^{-1})	Mean Annual Nutrient Load (kg ha^{-1})
	2003	2004	2003-2004	2004	2003	2004	2003-2004	2004
Vegetation								
Prairie	0.0022	0.0030	0.0171	0.0201	0.1172	0.0387	0.4656	0.5043
Turfgrass	0.0008	0.0009	0.0460	0.0469	0.0514	0.0080	0.4114	0.4194
P-value	0.13	0.05	0.38	0.41	<0.01	0.01	0.70	0.55

[a] Plots measured 2.44 m in width by 9.75 m in length or 23.78 m^2 and consisted of impervious:pervious ratios of concrete to prairie or turfgrass.

[b] Plots were dormant seeded Nov. 2002 and covered with wood-fiber mulch blankets. No vegetation was present during winter 2002-2003.

P values in this study followed very closely the soluble P values in quantity and percentage. In most studies, the bioavailable P quantities are greater than the soluble P fraction, but our BAP values averaged 91-93% of mean annual soluble P loading (Table II). Since the bioavailable P test was designed to quantify the amount of P contribution from sediment, the decreased results in the current study may be a function of too little sediment contained within runoff samples.

A common trend in the current study was greater P loading from unfertilized prairie vegetation as compared to fertilized turfgrass vegetation during the growing season. The quantities of P in runoff from prairie environments may be threshold standards to expect from unmanaged landscapes. Since the turfgrass was managed in this study, it is possible that the standard turfgrass practices of mowing and fertilization may have increased turf density, leading to decreased runoff nutrient loads. As was often the case, nutrient concentrations were higher from turfgrass than prairie vegetation but a decreased runoff volume moving through the turfgrass during the growing season may have caused decreased nutrient loading.

An interesting characteristic was that the final Autumn turfgrass fertility application in 2003 occurred on 6 November. Greater than 97% of the annual 2004 total P runoff occurred during the 2003-2004 winter with the first runoff event occurring on 20 February 2004. This was the first of 11 runoff events due to snowmelt or precipitation upon frozen soil, and occurred 106 days after the final 2003 fertility treatment. Once dissolved, fertilizer P applications become part of the soil P pool. The thin ice layer that separates the soil surface from late Winter and early Spring runoff waters may effectively isolate the fertilizer P and soil P pools from overhead runoff waters, yet large percentages of P were running off both turfgrass and prairie landscapes during this time period. Sediment sources may be due to sediment movement underneath the ice sheeting or due to variability in ice thickness. This may suggest that aboveground factors such as leaching from vegetation may be responsible for a large portion of the P pollution occurring in surface water runoff.

Conclusions

The current study is the first to report on total runoff volumes from both newly established turf and prairie vegetation under natural precipitation conditions. A ratio of at least 2:1 pervious:impervious surface was needed to significantly reduce runoff and contaminants. The fertilized turfgrass buffer strips allowed less total runoff volume, sediment and nutrients than the non-fertilized prairie buffer strip during the growing season. However, the total annual runoff volume, nutrient load and sediment load were similar due to such a large percentage of runoff occurring over frozen soil. Despite differences in the level of management, similar nutrient loading occurred from both vegetation types. Both current and future legislation and regulation need to account for ambient levels of nutrients exiting urban environments when delineating nutrient thresholds. Future research needs to further identify the impacts of vegetation density on water and nutrient usage within urban landscapes.

Acknowledgements

Funding for this project was supplied by Federal Hatch funds, project 5232.

References

1. Gordon, N.D.; McMahon, T.A.; Finlayson, B.L. *Stream Hydrology: An Introduction for Ecologists.* 1992. Baffins Lane; Chichester, West Sussex; John Wiley and Sons Ltd.
2. Leopold, L.B. *A View of the River.* 1994, Cambridge, MA, Harvard Univ. Press.
3. Hirsch, R.M.; Walker, J.F.; Day, J.C.; Kallio, R. *The Influence of Man on Hydrologic Systems.* In *Surface Water Hydrology.* Eds. W.G. Wolman and H.C. Riggs. Geological Society of America, 1990, Vol. 0-1.
4. Wisconsin Department of Natural Resources. *NR 151 "Runoff Management".* 2001. DNR-WT/2, Madison, WI. p 66.
5. Correll, D.L. *J. Environ. Qual.* **1998**, *27*, 261-266.
6. Erickson, J.E.; Cisar, J.L.; Snyder, G.H.; Volin, J.C. *Crop Sci.* **2005**, *45*, 546-552.
7. Miltner, E.D.; Branham, B.E.; Paul, E.A.; Rieke, P.E. *Crop Sci.* **1996**, *36*, 1427-1433.
8. Krenitsky, E.C.; Carroll, M.J.; Hill, R.L.; Krouse, J.M. *Crop Sci.* **1998**, *38*, 1042-1046.
9. Linde, D.T.; Watschke, T.L.; Jarrett, A.R.; Borger, J.A. *Agron. J.* **1995**, *87*, 176-182.
10. Timmons, D. R.; Holt, R.F. *J. Environ. Qual.* **1977**, *6(4)*, 369-373.
11. Brakensiek. D.L.; Osburn, H.B.; Rawls, W.J. *Field Manual for Research in Agricultural Hydrology.* USDA Agric. Handb. 224. 1979. U.S. Gov. Print. Office, Washington, D.C.
12. Stier, J.C.; Walston, A.; Williamson, R.C. *Int. Turf Soc. Res. J.* **2005**, *10*, 136-143.
13. Prairie Nursery. 2005. *Prairie Management Procedures* [Online]. Available at http://www.prairienursery.com/ (verified 22 Jul. 2007).
14. American Public Health Association. *Methods of Chemical Analysis of Water and Wastes.* In *Standard Methods for the Examination of Water and Wastewater.* Ed. M.A.H. Franson. USEPA. 1979, p. 444-450.
15. Murphy, J.; Riley, J.P. *Anal. Chim. Acta.* **1962**, *27*, 31-36.
16. Sharpley, A.N. *J. Environ. Qual.* **1993**, *22*, 678-680.

17. Mickelson, S.K.; Baker, J.L.; Ahmed, S.I. *J. Soil Water Cons.* **2003**, *58(6)*, 359-367.
18. Lush, W.M. *Agron. J.* **1990**, *82*, 505-511.
19. White, E. M.; Williamson, E.J. *J. Environ. Qual.* **1973**, *2(4)*, 453-455.
20. Kussow, W.R. In *The Fate of Nutrients and Pesticides in the Urban Environment.* ACS Symp. Ser. Vol 872. Nett, M.; Carroll, M.J.; Horgan, B.P.; Petrovic, A.M. Eds., 2008, *In Press.*
21. Easton, Z. M.; Petrovic, A.M. *J. Environ. Qual.* **2004**, *33*, 645-655.
22. Gross, C.M.; Angle, J.S.; Welterlen, M.S. *J. Environ. Qual.* **1990**, *19*, 663-668.

Chapter 10

Evaluation of Resource-Efficient Landscape Systems to Reduce Contaminants in Urban Runoff

James A. Reinert, B. Hipp, S. J. Maranz, and M. C. Engelke

Texas Agricultural Experiment Station, Texas A&M University Research and Extension Center, Dallas, TX 75252–6599

Urban non-point source pollution and the high cost of urban water treatment have generated interest to identify sources of pollution and to develop cost-saving preventive measures. A series of micro-landscape systems planted with either conventional vegetation or Resource Efficient Plants (REP^{TM}) was constructed in Dallas, TX to assess the contribution of stormwater runoff from residential landscapes into local non-point source pollution, and to quantify the impact of alternative best management practices. The conventional landscapes, receiving high inputs of fertilizer, herbicide and irrigation, produced significantly higher levels of nitrate nitrogen, orthophosphate and 2, 4-D in runoff. Stormwater runoff volume was also significantly higher under irrigated conditions. In contrast, the REP^{TM} landscape systems had lower levels of stormwater runoff and the concentration of nitrate nitrogen and orthophosphate in runoff was consistently very low, suggesting that fertilizer usage in residential landscapes poses little water quality risk. Levels of 2, 4-D in runoff were below federal drinking water thresholds. Reduced irrigation and chemical inputs made possible by REP^{TM} significantly reduced this risk. Conventional landscapes scored higher for visual ratings throughout the 3 year study period than did alternative REP^{TM} landscapes, indicating that drought resistant plants of both turfgrass and ornamental plants with better appearance are needed for widespread public acceptance of REP^{TM} landscape systems.

Increased public concern about water quality has generated considerable interest in identifying sources of pollution and developing methods to improve water quality, particularly in urban areas where large population concentrations may strongly impact water resources (1, 2, 3, 4). Although agricultural sources of non-point pollution are far more significant on a national scale than urban runoff (5), urban runoff can have a significant localized impact on water quality. In these areas, the construction of roads, parking lots and buildings creates extensive areas impervious to rainfall and results in increased volumes of runoff during storm events (6, 7, 8). The most recent National Water Quality Inventory reported that runoff from urban areas is the leading source of impairments to surveyed estuaries and the third largest source of water quality impairments to surveyed lakes (9). Additionally, populations and developmental trends indicate that by 2010, more than half of the Nation will live in coastal towns and cities and the runoff from these rapidly growing urban areas will continue to degrade our waters (10). Data from a number of research studies indicate that relatively little runoff occurs from turf surfaces due to normal irrigation and precipitation (11, 12, 13, 14). Depending on the magnitude of changes to the land surface, the total runoff volume can increase drastically. The North Texas Council of Governments estimates that 10-20% of rainfall will run off agricultural land, 40-50% will runoff residential areas and >90% will runoff the commercial urban areas. These changes not only increase the total volume of runoff, but also accelerate the rate at which runoff flows from the area (15). This increased volume of runoff, in turn, increases the chances that chemicals from various non-point sources will be carried into rivers, lakes or reservoirs, with potentially adverse effects on water quality (14, 16, 17, 18, 19, 20).

Within urban areas, different land use sectors vary in the amount and types of contaminants they contribute to receiving waters during rain storms. Commercial and industrial zones, especially automobile service stations, industrial storage and parking areas are often the most significant sources of organic toxicants and some metals (21). Residential landscapes have been implicated as contributors to some of the chemicals detected in water supplies (22). A survey of storm water discharge sites in the Dallas-Ft. Worth metroplex by the U.S. Geological Survey found that residential watersheds exceeded commercial and industrial watersheds in the production of seven pollutant constituents, including phosphates, chlordane and diazinon (23).

The turfgrass industry in the United States has an estimated value of $40 billion annually and covers 20.2 million ha. (50 million acres). Turfgrass is enjoyed by millions of homeowners with lawns, millions of athletes of all ages that use the sports fields, 26 million golfers and the many people that use parks and other recreational areas on a daily basis (24). The increasing emphasis by homeowners on having attractive, well maintained landscapes is attested to by the tremendous growth of the landscape industry (25, 26, 27). Conventional landscapes, especially under hot summer conditions like those annually in Texas, require frequent irrigation, as well as fertilizers, herbicides and other pesticides. Alternative landscapes comprised of Resource Efficient Plants (REP[TM]) requiring less water, chemical inputs and management practices, have been investigated in Texas (*28, 29, 30, 31, 32, 33, 34*) and other southwestern U.S.A. states (*35, 36, 37, 38*). Most of these efforts have been directed toward reducing water use, tolerance to low quality water, using more native plant materials and developing new plant materials better adapted to native environments. The potential impact of these REP[TM] on urban water quality has not received much attention.

The objectives of this study were to quantify fertilizer and herbicide losses in water runoff from conventional landscapes, assess the effectiveness of alternative REP[TM] landscapes in reducing runoff and runoff-borne contaminants, and to evaluate the potential acceptability of alternative landscapes to the public.

Materials and Methods

A series of 20 micro-landscapes, each 3.05 x 4.27 m or 14 m^2 (150 ft^2) in area, was constructed during 1989-1991 at the Texas A&M University Research and Extension Center at Dallas, TX, U.S.A. (Figure 1). The facility was built on undisturbed Austin silty clay soil with ca. 2% slope throughout the test area. Austin silty clay in the test area had a pH range of 7.8-8.2 with a naturally occurring, plant available concentration of phosphorus (range 5-12 mg/kg) in the upper 14 cm. Each plot was bordered with metal landscape edging (7.5 cm above soil line) and the down-slope end included a specially constructed aluminum sheet metal border that was inserted into the soil extending 20 cm below ground level. The lip of the metal border allowed runoff water to collect in a gutter system and drain into an 885 ml (calibrated at 885 ± 12 ml) electronic rain gauge. Each rain gauge was equipped with a magnetic reed switch to count each tip and was also provided with a 5.0 ml glass sample splitter tube that filled with water as it filled and would then dump into a funnel below that was connected by tubing to a series of solenoids. Sample collection was 5.0 ± 0.66 ml per tipping bucket. After a specified volume had been measured and the sample collected in a bottle, the solenoid closed and another one would open, thus diverting the sample to a new collection bottle. The system allowed for up to four individual subsamples before the remainder of a large runoff event was diverted into a final 3.5 liter bottle. The tipping bucket, solenoids, sample bottles and other collecting apparatus were all contained within a 900 liter stock tank covered with a tarpaulin to protect the apparatus from rain and other environmental elements. Runoff water samples from each runoff event greater than 3.18 mm (0.125 in.) were collected in 500 cc polyethylene bottles. Each of the 20 landscapes had identical measurements and collection systems and ca 2 %

Figure 1. Micro-landscapes, each 14 m² with 2% slope, under four management systems: A – Xeriscape; B – Low; C – Medium; D – High (runoff flow from front to back at the collection site). (See page 2 of color inserts.)

slope from front toward the collecting gutter. Runoff water that had been measured and sampled along with water from the surrounding area was removed through a drainage system at the lower end of the plots.

For each of the micro-landscapes, two-thirds of the surface area was planted with turfgrass and one-third was planted with shrubs. The whole shrubbery area was mulched with pine bark nuggets. Four management systems were investigated, each system was replicated five times and the replicates blocked according to the infiltration rate of the soil within each plot. Infiltration rate for each plot was determined by applying an irrigation event long enough to cause runoff from all plots. In this way, the five plots with the highest infiltration rates were grouped together and those with the lowest infiltration rate were grouped into the last replicate. Once plot identifications were assigned to the respective replicates, the plant materials could be installed for the respective treatments:

1. Xeriscape. Plots consisted of 'Prairie' buffalograss [*Buchloe dactyloides* (Nutt.) Engelm.], autumn sage [*Salvia greggii* (A. Gray)], and southern wax myrtle [*Myrica cerifera* (L.)], with no irrigation, fertilizer or herbicide applied during the experimental period.

2. Low Maintenance. Plots consisted of the same micro-landscape plants as the Xeriscape, but each year they received 98 kg nitrogen (N) ha^{-1} yr^{-1} and 24 kg phosphorus (P) ha^{-1} yr^{-1} applied in two applications (usually on 15 April and 15 September), and only 37 g of 2, 4-D ha^{-1} yr^{-1} applied as spot treatments to individual weeds or weedy areas (usually on 15 April) and with irrigation provided weekly to replace 25% of pan evaporation.

3. Medium Maintenance. Plots consisted of 'Tifgreen' bermudagrass [*Cynodon dactylon* (L.) Pers. x *C. transvaalensis* (Burtt-Davy)], 'Nana' dwarf yaupon holly [*Ilex vomitoria* (Ait.)], and eastern wax myrtle [*Myrica heterophylla* (Raf.)], with annual treatments of 292 kg N ha^{-1} yr^{-1} and 73 kg P ha^{-1} yr^{-1} applied in six applications (usually on the 15th of April, May, June, July, August and September) and 187 g of 2, 4-D ha^{-1} yr^{-1} applied in one spray application (usually on 15 April) and with irrigation provided weekly at 50% of pan evaporation.

4. High Maintenance. Plots consisted of the same plants as the Medium maintenance, but with annual treatments of 439 kg N ha^{-1} yr^{-1}, and 108 kg P ha^{-1} yr^{-1} applied in six applications (same dates as Medium) and 373 g of 2, 4-D ha^{-1} yr^{-1} applied in two spray applications (usually on 15 April and 15 September) and with irrigation provided weekly at 70% of pan evaporation.

Runoff from each micro-landscape was collected and sampled immediately after, or the next morning after, each rain event and removed to the laboratory for analysis. Fractional samples taken during each runoff event were analyzed for nitrogen, phosphorus and 2, 4-D content. The volume of runoff was calculated using the number of tips from the tipping rain gauge.

Samples were evaluated for nitrate-N using Orion Ion Specific Procedures with an Orion Nitrate 9707BN Electrode (Thermo Electron, Water Analysis/Orion Products, 166 Cummings Center, Beverly, MA 01915-6199). The amount of phosphorus in each sample was determined by Hack procedures 365.2 (Hack Co., P.O. Box 389, Loveland, CO 80539) using a Baush & Lomb Micro-Flo Three Spectronic 100 (Baush & Lomb, 820 Linden Ave, Rochester, NY 14625) (*39, 40*). The amount of 2, 4-D in samples was assayed by immunoassay procedure using a 2, 4-D Ohmicron RaPID Assay® Kit (Ohmicron Environmental Diagnostics, Inc., 375 Pheasant Run, Newtown, PA 18940).

The turf area of each plot was mowed weekly, with clippings returned during the growing season. Buffalograss was mowed at 5 cm and bermudagrass was mowed at 2.5 cm height of cut. The entire plot was rated monthly for overall appearance. For each landscape plot, the turfgrass and shrubs were given a combined visual appearance rating (1-10; where 10 indicated lush dark green plants and 1 represented poor quality brown vegetation).

Data were analyzed using analysis of variance procedures (1-way ANOVA), however, when REP^{TM} landscapes vs. conventional landscapes were compared, a split-plot analysis was performed. Means were separated by Fisher protected

170

least significant differences (LSD) (P = 0.05) using CoStat statistical software (*41, 42*).

Results

Data from the micro-landscapes was collected from late 1993 through 1996. During the monitoring period there were two storms that produced runoff in 1993, eleven in 1994, eight in 1995 and one in 1996. An extended drought from late 1995 through 1996 limited the runoff collection to just one storm (28 October 1996) during the dry period. Total monthly rainfall, including these rain events during the test period, is presented in Figure 2. The 1994 and 1995 seasons yielded the most complete data sets. Runoff by management level for these two years is given in Figures 3 and 4.

Differences in runoff volume between treatments varied seasonally. During the cool-wet seasons (15 November to 15 May) for 1993-1996, which corresponds to the period of greater soil moisture, no significant difference occurred in the degree of runoff between the lower (REP^{TM}) and higher (conventional) maintenance landscapes. In contrast, during the warm-dry season (mid-May to mid-November), the more heavily irrigated conventional landscapes yielded significantly greater runoff (160%) than the REP^{TM} landscapes (Figure 5).

However, examining the different irrigation management levels within plant communities indicates that 'any amount of irrigation' will impact the level of runoff experienced. As more water is retained in the soil profile, the infiltration capacity is reduced for the landscape, which then results in greater runoff (Figure 6). The actual runoff is however a very low percentage of the total rainfall received over each year of the study. This pattern of runoff is further supported by the increased runoff experienced during the warm-dry seasons (1994-95) as it relates to higher irrigation (management) levels (Figure 7). Water demands for the Low management buffalograss were much lower than for the bermudagrass in the Medium and High management plots; this difference resulted in a significantly greater amount of runoff from the Low management plots during the cool-wet seasons when more moisture was available in the soil profile from rainfall (Figure 6).

Rainfall for the project monitoring period totaled 3,290 mm (132 in). The microlandscapes under high irrigation produced a mean of 2,540 m^3 ha^{-1} of runoff from storms during this period, representing a 7.6% runoff rate. The mean runoff rate for the non-irrigated Xeriscapes was 4.4%, ca. one-half as much. This runoff rate was a relatively low percentage of the total rainfall, especially when compared to estimates by the North Texas Council of Government of 40-50% runoff from residential areas (*15*). To put these estimates into perspective, however, their estimates included the entire residential area including turf

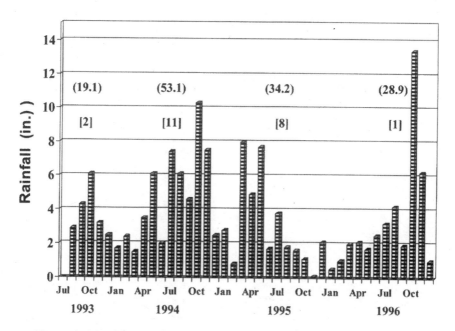

Figure 2. Monthly rainfall (total inches per year) on the microlandscapes during the monitoring period (1993-1996).

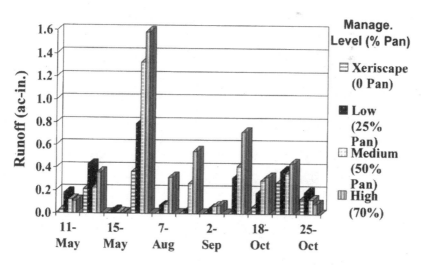

Figure 3. Acre-inches of runoff from management levels on microlandscapes during the 1994 season.

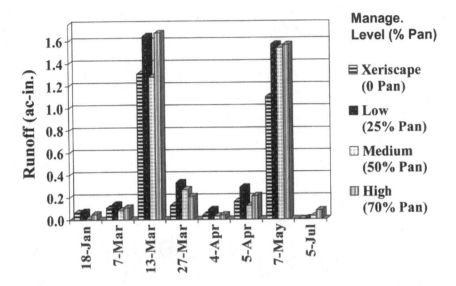

Figure 4. Acre-inches of runoff from management levels on microlandscapes during the 1995 season.

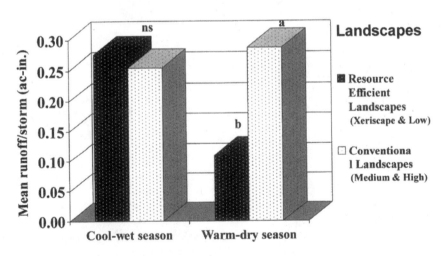

Figure 5. Seasonal differences (1993-1996) in runoff between Resource Efficient Landscapes and Conventional Landscapes from microlandscapes – Dallas, TX. Data means for a season with the same letter are not significantly different by Fisher's protected LSD (P = 0.05).

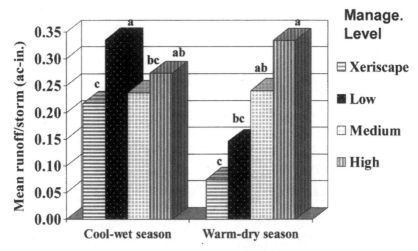

Figure 6. Seasonal differences (1993-1996) in runoff among management levels for microlandscapes during cool-wet vs. warm-dry seasons – Dallas TX. Data means for a season with the same letter are not significantly different by Fisher's protected LSD (P = 0.05).

landscape plants and hard surfaces, whereas the present study includes only the turf and landscape plant areas. Although the difference in overall runoff rates between landscapes receiving high irrigation and those receiving no irrigation may appear slight, runoff was less likely to occur from the Xeriscapes, particularly during the warm-dry months and especially if the time interval between rainfall events was longer. The 10 July 1994 rain event, following nearly 2-months of no runoff (over 50 days of sparse rainfall), clearly demonstrates this concept (Figure 3), whereas the frequency of runoff events in the Fall increased with more frequent rains and a higher saturated soil profile. A lone runoff event in the 1996 drought year (on 28 October, following over 16 months of insignificant rainfall) produced runoff from all five replicates of the High and Medium landscapes, while only three of the Low and one Xeriscape plot produced runoff.

Rainfall patterns are obviously seasonal, but also vary from year to year, all of which impacts irrigation, fertilizer and pesticide usage. Annual rainfall over this 4 year period averaged 1,092 mm (43 in.), with a high of 1,348.7 mm (53 in.) in 1994 to a low of 868.7 mm (23 in.) in 1995. The 50-year average for the Dallas/Fort Worth Metroplex is 940 mm (37 in.). The 1994 runoff data (Figure 7) reflect that year's high rainfall pattern, which mostly occurred in the warmer season from July through October (Figure 3). Conversely, the 1995 runoff data reflects the low but more pronounced rainfall pattern during the cooler season of 1995 (Figure 4). The last significant rainfall in 1995 occurred on 5 July.

The warm season grasses used in this study will typically go dormant from mid-December until early-March. The level of N and P runoff during the

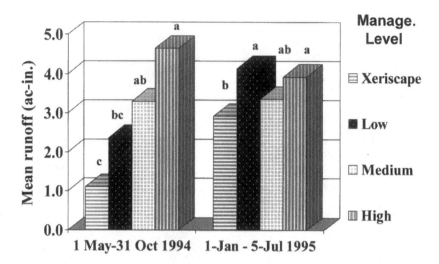

Figure 7. Total acre-inches of runoff from management levels during the 1994-1995 monitoring period – Dallas TX. Data means for a season with the same letter are not significantly different by Fisher's protected LSD (P = 0.05).

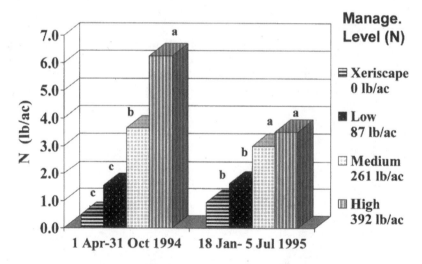

Figure 8. Total loss of nitrate-nitrogen in runoff from management levels during the 1994 and 1995 monitoring periods – Dallas TX. Data means for a season with the same letter are not significantly different by Fisher's protected LSD (P = 0.05).

growing seasons in 1994 and 1995 (Figures 8 and 9) appears highly correlated to the irrigation and pesticide application levels. Additionally, the level of N and P in the runoff correlates closely with the actual runoff from each management level for each year. The reduced total amount during 1995 reflects the lesser rainfall that year. In reality, the greater the application rates of chemicals, the greater the potential loss.

The highest mean annual loss of nitrate nitrogen occurred on the High input landscapes [7.2 kg NO_3-N/ha (6.3 lb NO_3-N/ac)] in 1994, from the application rate of 439 kg NO_3-N/ha (392 lb NO_3-N/ac). However, this loss represents only 1.63% of the applied nitrogen. In terms of water quality, the mean concentration of NO_3-N in runoff from the High input landscapes was 4.4 mg L^{-1} (ppm) for the project period (Figure 10), well below the USEPA drinking water standard of 10 mg L^{-1} (ppm) for nitrogen (43). Only one storm produced runoff with a mean nitrogen concentration [10.8 mg L^{-1} (ppm)] that exceeded the safe drinking limit. The elevated N level in this event can be explained readily, since this storm was the first rain event which also produced runoff following the fertilizer applications ca. 3 weeks earlier on 15 September.

For phosphorus, the highest mean annual loss, occurring from the High input landscapes in 1994 (Figure 9), was 1.0 kg P/ha (0.9 lb P/ac), from the application rate of 108 kg P/ha (97 lb P/ac). This loss represented less than 1% of the applied phosphorus. The mean concentration of phosphorus in storm runoff for the High input landscapes was less than 1 mg L^{-1} (1 ppm) for the project period (Figure 10).

Total loss of 2, 4-D in runoff was significantly higher during 1994 than during 1995 (Figure 11). The highest mean annual loss, occurring on the High input plots in 1994, was 3.7 g of 2, 4-D ha^{-1} (1.5 g of 2, 4-D/ac), from an application rate of 373 g 2, 4-D ha^{-1} (151 g of 2, 4-D/ac); and represented a loss of 1% of applied pesticide The concentration of 2, 4-D in runoff from the high management landscapes averaged 0.003 mg L^{-1} (ppm) for the project period (Figure 12). This is less than half the USEPA drinking water standard of 0.007 mg L^{-1} (ppm) for 2, 4-D (40). The 2, 4-D concentration in runoff following most storms was very low, but the storm runoff on 7 October 1994 did exceed USEPA standards with a 2, 4-D spike of 0.018 mg L^{-1} (ppm) and the concentration was uniform across all 5 replications. The Fall application of 2, 4-D had been applied only 18 days earlier and the plots had received only 1.3 mm of rainfall before the 7 October runoff event. This one time occurrence points out how important timing of pesticide applications is in relation to irrigation practices and rainfall events.

The High management landscapes consistently looked best in monthly visual ratings, but were not always significantly better than the Medium or Low management plots. The yearly mean for High management landscapes was significantly better than for the Medium in 1994, but not significantly better in 1995 (Figures 13 and 14). Depending upon the time of year, the Low

176

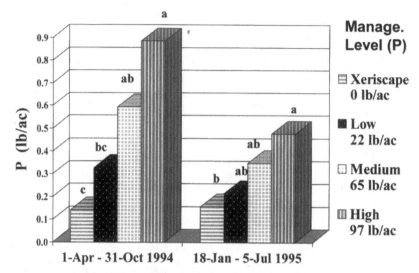

Figure 9. Loss of phosphorus (P) as orthophosphate in runoff from management levels during the 1994 and 1995 monitoring periods – Dallas TX. Data means for a season with the same letter are not significantly different by Fisher's protected LSD (P = 0.05).

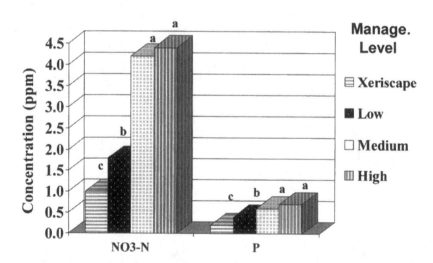

Figure 10. Mean chemical concentration of nitrate-nitrogen and phosphorus (Orthophosphate) per storm runoff event from management systems (1993-1996) – Dallas TX. Data means for a season with the same letter are not significantly different by Fisher's protected LSD (P = 0.05).

management plots ranked sometimes better but sometimes worse than the Medium plots. The Xeriscape, with no irrigation, fertilizer or herbicides applied, almost always ranked lowest. All four management levels are pictured in Figure 15 during the Summer stress period. During the study period, the buffalograss thinned with noticeable weed encroachment in the Xeriscape. The bermudagrass sustained white-grub damage requiring insecticidal treatment. Some weed colonization occurred in the bare areas of the bermudagrass turf caused by grub damage. No white-grub damage was observed in the Low or Xeriscape buffalograss turf.

None of the southern wax myrtles (*Myrica cerifera*) selected as dwarf trees for the Xeriscape and Low management plots survived; most were replaced twice. The autumn sage (*Salvia greggii*) shrubs survived and provided good flower color throughout the study under both Low and Xeriscape management.

Discussion

Dallas has a bimodal precipitation pattern, with peak rainfall in the Spring and Fall. High daily temperatures in early Summer quickly dry out residual soil moisture from Spring rains. Fall precipitation gradually recharges soil moisture depleted during the hot, dry Summer. During the study period, an annual shift in runoff pattern indicating significant soil dry-down was observed in mid-May, while another shift indicating full soil moisture recharge was observed in mid-November.

During the warm-dry period of the year, substantially less runoff occurred from landscapes with REP^{TM} than from the more conventional landscapes. This difference can be attributed to lower soil moisture under the buffalograss due to lower irrigation levels. Reduced irrigation on the buffalograss/salvia vegetation in the Low management landscapes and Xeriscapes was made possible by the greater tolerance of these plants to low soil moisture. The landscapes with REP^{TM} were able to capture and hold precious and often sparse summer rains, reducing runoff and consequently reducing the need for irrigation. Chemical losses were also lower in the REP^{TM} landscapes, due both to less runoff and lower chemical application rates. Conventional landscapes typically have higher soil moisture content and therefore are prone to greater runoff potential and chemical loss. The greatest loss of fertilizer and pesticides occurred in the runoff from conventional landscapes under irrigated conditions during the warm-dry season when most fertilizers and pesticides are applied.

During the cool-wet season, more runoff occurred on the buffalograss plots than on the bermudagrass plots, although the difference was not significant (Figure 6). Since irrigation during this season was seldom required due to low pan evaporation rates, soil moisture differences across treatments were minor.

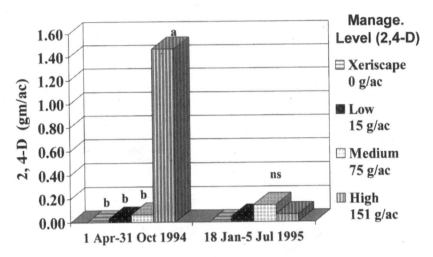

Figure 11. Loss of 2,4-D in runoff from management levels during the 1994 and 1995 monitoring periods – Dallas TX. Data means for a season with the same letter are not significantly different by Fisher's protected LSD (P = 0.05).

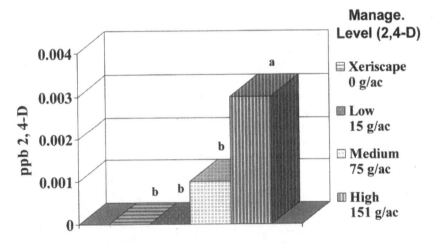

Figure 12. Mean concentration of 2, 4-D per storm runoff (1993-1996) from microlandscapes – Dallas TX. Data means for a season with the same letter are not significantly different by Fisher's protected LSD (P = 0.05).

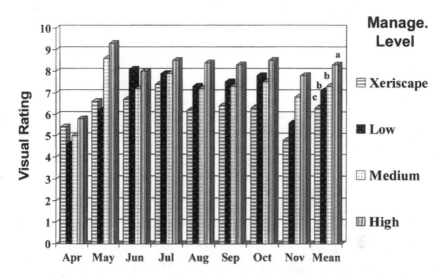

Figure 13. Monthly visual rating of microlandscapes under management levels during 1994 (10 = dark green, dense: 1 =brown, poor quality).

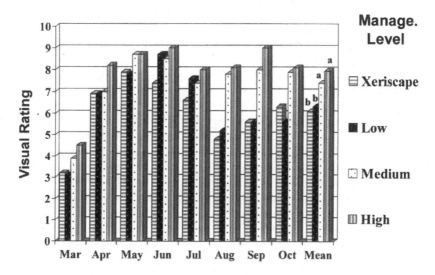

Figure 14. Monthly visual rating of microlandscapes under management levels during 1995 (10 = dark green, dense: 1= brown, poor quality).

Figure 15. Appearance of four management systems during summer stress (warm-dry period). (See page 2 of color inserts.)

The slight difference in runoff rate under relatively uniform soil moisture conditions appears to be due to differences in turfgrass growth-form. Linde et al (*13*) compared hydrologic characteristics of perennial ryegrass with creeping bentgrass and found that less runoff occurred from the bentgrass. Runoff from ryegrass was also initiated more quickly. Creeping bentgrass (*Agrostus palustris* Huds.) has more tillers per unit of area and produced a thatch layer, both of which contributed to increased interception, slower water movement and greater infiltration. An additional study compared nitrogen runoff and leaching between newly established St. Augustinegrass turf and an alternative residential landscape of mixed grasses (*44*). Through the first year following installation of the landscapes, fertilizer N loss in surface runoff was insignificant. In contrast, N leaching losses were significantly greater on the mixed-species landscapes compared with the St. Augustinegrass, annually. Their results indicated that St. Augustinegrass [*Stenotaphrum secundatum* (Walt.) Kuntze] was more efficient at using applied N, and minimizing N leaching, compared with the alternative landscape. The cool-season runoff pattern measured in our study appears to reflect similar differences in growth form between the grasses we used. The bermudagrass is far denser and forms a substantial water holding mat, including the thatch layer, while the buffalograss is thinner, with little thatch and litter.

The dense growth-form of bermudagrass plays a role in the low nitrate and phosphate losses measured throughout the year (Figure 9). In spite of the 439 kg N ha^{-1} application rate on the high management landscapes, nitrate losses in runoff were less than 2% of applied nitrogen. Phosphate losses were less than 1%. In terms of water quality, the mean NO_3-N concentration of 4.4 mg L^{-1} in runoff from the High input landscapes was well below the USEPA drinking water standard of 10 mg L^{-1} for nitrate (*43*). The highest nitrate concentration measured in runoff for an individual storm was 10.8 mg L^{-1}. This indicates that runoff coming directly from highly fertilized landscapes was clean even before further dilution with much greater volumes of stormwater flowing from impervious surfaces. Other studies have shown a similar ability of a turfgrass cover to reduce runoff, and therefore to enhance the soil water infiltration and ground water recharge potential (*14, 45, 46, 47, 48*). Most runoff from urban landscapes appears to develop from impervious surfaces (*6, 7*). Finally, the reduced runoff volume from a good turf cover offers the potential to decrease the stormwater management requirements and costly structures used in urban development (*49*).

An approximate dilution factor for turfgrass runoff constituents in Dallas-Ft. Worth Metroplex residential stormwater discharges can be calculated by combining our small plot data with U.S. Geological Survey findings (USGS). In a study of 11 metroplex residential watersheds, the USGS measured a 27% mean runoff rate (*50*). In contrast, Burton and Pitt showed urban runoff accounted for ca 10% of nonpoint surface pollution (*51*). Using our finding of a 7.6% storm runoff percentage from irrigated bermudagrass on Austin silty clay soil, a 1:3 dilution ratio can be approximated.

The USGS measured a mean $NO_3+ NO_2$-N concentration of 0.6 mg L^{-1} at 11 residential storm water discharge sites in the metroplex, with a high measurement of only 1.7 mg L^{-1}. When the 1:3 dilution ratios are factored in, the USGS figures closely correspond to our data. These measurements indicate that fertilizer losses from residential landscapes in the Dallas-Ft. Worth metroplex do not pose a significant water quality threat.

Other studies have also reported that the nitrogen content of runoff from areas under turfgrass is generally quite low (*17*), and usually below federal drinking water standards (*48, 52*). Although nitrogen losses from turfgrass runoff do not appear to be a cause for concern, there is a greater risk of loss if a runoff event occurs soon after application of the fertilizer (*17, 31, 52, 53, 54, 55*). McLeod & Hegg (*53, 56*) found the nitrate concentration in runoff exceeded water quality standards only during the first runoff event after application. Where infiltration rates are high, fertilizer losses from turf runoff have been found to be very low (*57*).

In contrast to low levels of nitrogen and phosphorus measured in turfgrass runoff, chemicals that are hazardous at much lower concentrations, such as 2, 4-D and many insecticides, appear to pose a greater water quality hazard. In one runoff event (7 October 1994), a 0.018 mg L^{-1} concentration of 2, 4-D was measured (Figure 12). When the 1:3 stormwater dilution rates is factored in, the estimated 2, 4-D concentration in stormwater discharge would likely be below the federal drinking water threshold of 0.007 mg L^{-1} (43). However, the data suggest that a heavier 2, 4-D application could have resulted in toxic levels of this chemical in storm water discharge. Other commonly applied hazardous lawn chemicals should be cause for greater concern.

The storm referred to above occurred after a month with no runoff and was the first runoff event following the fall 2, 4-D application on the high management plots. During the same storm, runoff from the low management plots was negligible while no runoff occurred from the Xeriscapes. This exemplifies the potential risk from highly irrigated landscapes in combination with applications of herbicides and insecticides toxic at very low concentrations. Whereas runoff should have been minimal after a dry month, the high soil moisture level needed to maintain the Tifgreen bermudagrass turf resulted in runoff with a higher but not excessive loss of the 2, 4-D applied.

Our experience shows that plants must be carefully chosen for the level of management they will receive. *Salvia greggii* was a good choice for both low management and Xeriscape landscapes, but *Myrica cerifera* did not survive with reduced irrigation. The buffalograss provided an acceptable turf under both of the reduced management levels, but usually ranked better with the input provided under Low management. Water efficient plants with good insect and disease resistance and good competitive ability with weeds will further reduce the need for insecticides and herbicides. The selection of an alternate turfgrass species with similar drought resistance to buffalograss but with a denser growth-form may result in an even greater reduction in runoff and chemical losses than was provided by the resource efficient landscape used in this study.

Conclusions

Resource efficient landscapes can reduce pollution by reducing runoff potential through maintaining a drier soil profile. Heavily irrigated landscapes have a greater potential for runoff than do landscapes maintained in a drier state. Fertilizer application in residential landscapes poses little pollution risk. Pesticides that are hazardous at low concentrations, such as diazinon and 2, 4-D, have the potential to adversely impact urban water quality. The greatest chemical losses are registered when a storm occurs soon after chemical application in heavily irrigated landscapes.

Selecting plant materials with lower inherent water needs will impact the amount of water, fertilizer and pesticides used and potentially lost in runoff. The success of efforts to switch to Resource Efficient Plants is contingent upon consumer acceptance of alternative landscapes. In a California study of public response to resource efficient landscapes, xerophytic shrubs met with broad approval, whereas the public was less enthusiastic about turfgrass (tall fescue) under reduced irrigation--yet turfgrass was preferred over a xerophytic coyote brush ground cover (58). In our study, highly managed bermudagrass consistently outclassed low management buffalograss in appearance. Consumer preference for lush, green turf emphasized the need to select turfgrass varieties that form denser, greener stands under reduced irrigation and inputs. The availability of green, drought tolerant turfgrasses could be the key to popularizing resource efficient landscapes and implementing residential runoff and chemical reduction programs based on the REP^{TM} concept. The more recent development of new drought tolerant cultivars of bermudagrass, zoysiagrass (*Zoysia* spp.) and St. Augustinegrass with excellent turf quality may help satisfy the need for more durable turfgrass for resource efficient landscapes.

Acknowledgements

Partial support for this study was provided by a grant (Project # 8575001, Contract # 420000-0006) from the U.S. Environmental Protection Agency through the Texas Natural Resources Conservation Commission. Appreciation is extended to Mary Ann Hegemann, Dennis Hays, Tim Knowles, Tim Brubaker, Sue Metz and Jeffery Powell for technical assistance during the duration of this study.

References

1. Bundy, L.G.; Andraski, T.W.; Powell, J.M. *J. Environ. Qual.*, **2001**, *30*, 1822-1828.
2. Cisar, J.L.; Snyder, G.H. *Fate and Management of Turfgrass Chemicals*; Am. Chem. Soc., Washington, D.C., 2000.
3. Niedzialkowski, D.; Athayde, D. *Perspectives on Nonpoint Source Pollution*; Proc. National Conf. 19-20 May 1985, Kansas City, MO, 1985, p. 437-441.
4. Schoumans, O.F. a. P.G. *J. Environ. Qual.,* **2000**, *29*, 111-116.
5. Myers, C.F.; Meek, J.; Tuller, J. S.; Weinberg, A. *J. Soil Water Conserv.*, **1985**, *40*(1), 14-18.
6. Jensen, R. *Environ. Protection*, **1995**, *4*, 38-40.

7. Stier, J.C.; Walston, A; Williamson, R.C. *Int. Turfgrass Soc. Res. J.*, **2005**, *10*, 136-143.
8. Walker, W.J.; Balogh, J.C.; Kenna, M.P.; Snow, J.T. *U.S. Golf Assoc., Green Section Record,* **1990**, *28*, 7-8.
9. Anonymous, *Managing Urban Runoff. Polluted Runoff (Nonpoint Source Pollution)*U.S. Environ. Protection Agency, 2000, EPA No. 816R00013, 90 p.
10. Anonymous. *Managing Urban Runoff. Polluted Runoff (Nonpoint Source Pollution)*; U.S. Environ. Protection Agency, 2004, Pointer No. 7, EPA841-F-96-004G: 3 p. http://www.epa.gov/owow/nps/facts/point7.htm.
11. Kussow, W.R. *1995 Agron. Abstr., Am. Soc. Agron.,* Madison, WI, **1995**, p. 157-158.
12. Linde, D.T.; Watschke, T.L. *J. Environ. Qual.*, **1997**, *26*, 1248-1254.
13. Linde, D.T.; Watsche, T.L.; Jarrett, A.R.; Borger, J.A. *Agron. J.*, **1995**, *87*, 176-182.
14. Watsche, T.L.; Mumma, R.O. Pennsylvania State Univ., *Environ. Resources Res. Inst.*, 1989, ER-8904.
15. NCTCOG. *Integrated Storm Water Policy Guidebook*; North Central Texas Council of Governments. "Review Draft", Sep. 2004, 2004.
16. Brach, J. *Protecting Water Quality in Urban Areas-best Management Practices for Minnesota*; Minnesota Pollution Control Agency, Div. Water Quality, 1989.
17. Gross, C.M.; Angle, J.S.; Welterlen, M.S. J. E*nviron. Quality*, **1990**, *19*, 663-668.
18. Leonard, R.A. *In* R. Grower (ed.). *Environmental Chemistry of Herbicides*; CRC Press, Boca Raton, FL, 1988, Vol. 1, Chap. 3.
19. Petrovic, A.M.; Borromeo, N.R. In *Handbook of Integrated Pest Management for Turf and Ornamentals*; Leslie, A.R., Ed., Lewis Pub., Boca Raton, FL, 1994, p. 29-51.
20. Watsche, T.L.; Harrison, S.; Hamilton, G.W. *U.S. Golf Assoc. Green Section Record,* **1989**, *27*, 5-8.
21. Pitt, R.; Field, R.; Lalor, M.; Brown, M. *Water Eviron. Res.,* **1995**, *67*, 260-275.
22. Daniel, T.C.; Wendt, R.C.; Konrad, J.G. *Univ. Wisconsin Coop. Ext. Bull.*, 1978, G-2958. 8 p.
23. NCTCOG. North Central Texas Council of Governments. Jan. 12, 1993, Dallas, TX, 1993, Final *Summary Report – Task 2.0*.
24. Anonymous. National Turfgrass Federation, Inc., *National Turfgrass Evaluation Program*. Beltsville, MD, 2003, www.ntep.org/pdf/turfinitiative.pdf.
25. Funk, R. In *Integrated pest management for turfgrass and ornamentals*; A.R. Leslie. A.R.; Metcalf, R.L., Eds., U.S. Environ. Protection Agency PB90-204587, Washington, DC, **1989**, p. 97-105.

26. Watsche, T.L. *U.S. Golf Assoc., Green Section Record,* **1986**, *24*, 6-7.
27. Wilkinson, J.F. In *Advances in Turfgrass Entomology*; Niemczyk, H.D.; Joyner, B.G., Eds., Hammer Graphics Inc., Piqua, OH, **1982**, p. 141-142.
28. Duble, R.L.; Welsh, W.C. *Texscape for Conservation*; Texas Agric. Ext. Serv. Bull., 1985. B-1498.
29. Engelke, M.C. *Turfgrass Res. Rep. Summary*; U.S. Golf Assoc., Far Hill, NJ, 1993, p. 8-9.
30. Engelke, M.C.; Lehman, V.G. TX Agric. Exp Stn., 1990, *Leaf. L-2419.*
31. Hipp, B.; Alexander, S.; Knowles, T. *Water Sci. Tech.,* **1993**, *28*(3-5): 205-213.
32. Hipp, B.W.; Engelke, M.C.; Simpson, B.J. In *How healthy is the upper Trinity River? Biological and water quality perspectives*; Jensen, R., Ed.. Texas Water Resources Institute, Texas A&M Univ., College Station, TX, 1991, p. 183-190.
33. Simpson, B.J.; Hipp, B.W. In *Conf. Proc. Water for the 21st Century: Will it be There? Freshwater Soc. for the Center for Urban Water Studies*; Collins, M.A., Ed., S. Methodist Univ. April 3-5, 1984, Dallas, TX, **1984**, p. 483-500.
34. Simpson, B.J.; Hipp, B.W. In *Erosion Control - Protecting our Future*; Proc. Conf. XVII. Int. Erosion Control Assoc. 27-28 Feb. 1986, Dallas, TX, 1986, p. 141-151.
35. Cuany, R. In *1993 Turfgrass Res. Rep. Summary*; U.S. Golf Assoc., Far Hill, NJ, 1993, p. 20-21.
36. Johnson, E.A. *Grounds Maintenance*, **1989**, *189*(7), IR8-IR18.
37. Kneebone, W.R.; Pepper, I.L. *Agron. J.,* **1981**, *74*(3), 419-423.
38. Sacamano, C.M.; Jones, W.D. Univ. Arizona Coop. Ext. Serv., **1975**, *Bull. A-82.*
39. Anonymous. In *Part 2, Analysis Procedures. Procedures for Water and Wastewater Analysis*; Hach Co., Loveland, CO, 1987, 2nd Edit, p. 2-91- 2-92.
40. Greenberg, A.E.; Clesceri, L.S.; Eaton, A.D., Eds. In *Standard Methods for the Examination of Water and Wastewater*; Am. Public Health Assoc., Washington, DC, 1992, 18th Edit, p. 2-90 - 2-95.
41. CoHort Software. *CoHort Software, Manual*; Rev. 4.20. Berkeley, CA. 1990.
42. Sokal, R.R.; Rohlf, F.J. *Biometry*, W.H. Freeman and Co., San Francisco, CA, **1981**.
43. Pontius, F.W. *J. Am. Waste Water Assoc.,* **1995**, *87*, 48-58.
44. Erickson, J.E.; Cisar, J.L.; Volin, J.C.; Snyder, G.H. *Crop Sci.,* **2001**, *41*, 1889-1895.
45. Bennett, H.H. *Soil Conservation*; McGraw-Hill Book Co., New York, 1939.
46. Gross, C.M.; Angle, J.S.; Hill, R.L.; Welterlen, M.S. *J. Environ. Quality,* **1991**, *20*, 604-607.

47. Jean, S.; Juang, T. *J. Agric. Assoc. China*, **1979**, *105*, 57-66.
48. Morton, T.G.; Gold, A.J.; Sullivan, W.M. *J. Environ. Quality*, **1988**, *17*, 124-130.
49. Schuyler, T. *Controlling Urban Runoff: A Practical Manual for Planning and designing Urban BPMs*; Metropolitan Washington Council of Governments, Washington, DC, 1987.
50. NCTCOG. *Storm Water Discharge characterization, Final Summary Report*; *North Central Texas Council of City Governments, March 23, 1994, Dallas, TX*, 1994.
51. Burton, G.A. Jr.; Pitt, R.E. *Stormwater effects handbook, Lewis Publishers,* Boco Raton, FL, 2002.
52. Shuman, L.M. *J. Environ. Quality*, **2002**, *31*, 1710-1715.
56. McLeod, R.V.; Hegg, R.O. *J. Environ. Quality*, **1984**, *13*, 122-126.
57. Brown, K.W.; Thomas, J.C.; Duble, R.L. *Agron. J.* **1982**, *74*, 947-950.
58. Thayer, R.L. *HortScience,* **1982**, *17*, 562-565.

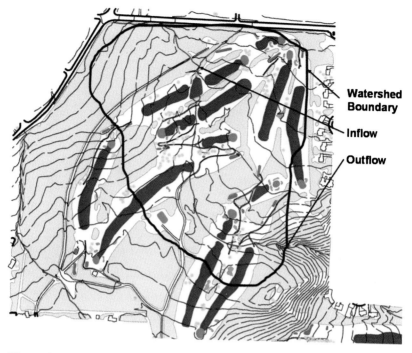

Figure 5.1. Layout of Northland Country Club Golf Course and study area.

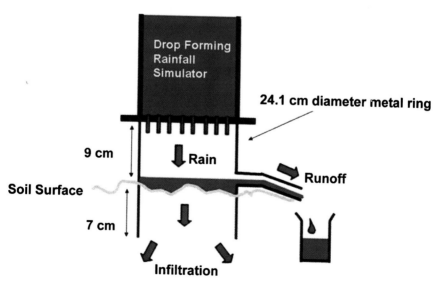

Figure 6.2. Diagram of rainfall simulator used in this study. More information available in Ogden et al (10).

Figure 10.1. Micro-landscapes, each 14 m² with 2% slope, under four management systems: A – Xeriscape; B – Low; C – Medium; D – High (runoff flow from front to back at the collection site).

Figure 10.15. Appearance of four management systems during summer stress (warm-dry period).

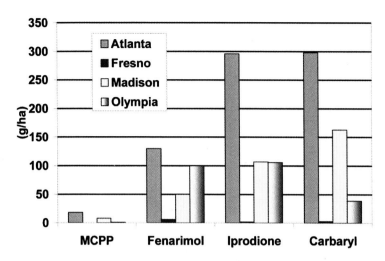

Figure 12.1. Mean annual runoff of four pesticides at four fairways sites.

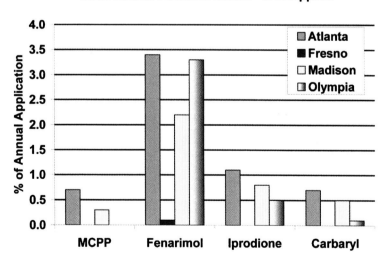

Figure 12.2. Mean annual fairway runoff pesticides expressed as a percentage of annual application.

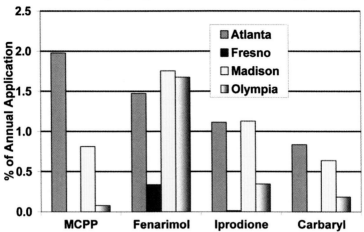

Figure 12.3. One in ten year pesticide runoff events.

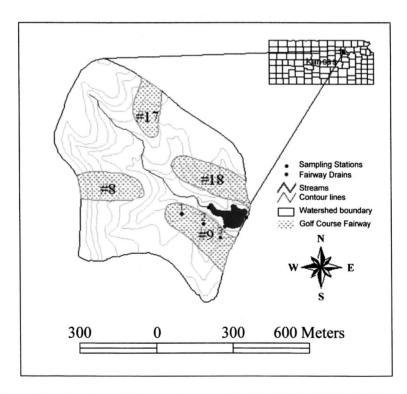

Figure 14.1. Location of fairway drains on hole #9 at Colbert Hills Golf Course. The fairway drains eventually connect and drain into the detention pond.

Chapter 11

Thatch Pesticide Sorption

Mark Carroll

Department of Plant Science and Landscape Architecture, University of Maryland, College Park, MD 20742

In turf that contains thatch, pesticide mobility is largely dependent on the sorption properties of the thatch. Sorption measurements of pesticides to thatch are relatively few in number. As a result, most thatch pesticide sorption coefficients used in computer modeling efforts are obtained from pesticide property databases. These databases largely consist of sorption measurements obtained from mineral soils. Calculation of the linear distribution coefficient (K_d) from the organic carbon partition coefficient (K_{oc}) using such databases will usually overestimate the sorption capacity of thatch. A comparison of the few studies that have measured pesticide sorption to thatch and the underlying soil suggests that thatch organic carbon has about 60% of the sorption efficiency of mineral soil organic carbon. It is proposed that improved estimates of pesticide sorption to thatch may be realized by considering the polarity of thatch organic matter.

The ability of computer simulation models to provide reliable estimates of the fate and transport of chemicals is highly dependent on the use of parameter values that accurately represent the crop and management conditions of interest. Sensitivity analysis of leaching models frequently reveals that pesticide leaching losses are most affected by the sorption and degradation parameter values used in the model simulation (*1*). Attempts to validate models for use in turf have demonstrated that model performance is highly sensitive to the sorption values assigned to thatch (*2*). This is consistent with several field and laboratory based investigations that indicate thatch is the primary attenuator of pesticide movement in a turf (*3-6*). Poor model performance in turf environments has been partially attributed to inadequate characterization of the sorption properties of turfgrass thatch (*7, 8*). This paper examines the current state of knowledge of pesticide sorption in turfgrass thatch and reviews common approaches used to measure or estimate sorption. Sorption is discussed in the context of mathematical modeling in recognition of the fact that modeling efforts represent one of the primary uses of sorption data.

Sorption parameters for modeling are typically obtained in one of three ways. The usual approaches are to: 1) obtain the required parameter or parameters by direct measurement, 2) to use consensus values derived from the literature, or to 3) use the chemical properties of the pesticide to estimate sorption. The first approach is most applicable to site specific evaluation of pesticide transport and is also widely used to evaluate model performance. Literature based sorption parameters are also often used in these same situations to either minimize the cost of the modeling effort, or because it is neither feasible nor practical to obtain direct measurements of pesticide sorption. For example, in the case of a computer modeling effort undertaken to develop a water quality risk assessment for a proposed golf course site, it is not possible to directly measure the sorption properties of a thatch layer for a turf area that does not yet exist. Estimating the sorption properties of a pesticide from its chemical and physical properties occurs when there is a dearth of direct measurements to arrive at a consensus sorption estimate. This is typically the case for many industrial chemicals and for pesticide metabolites (*9*).

Direct Measurement of Sorption

Direct measurement of pesticide sorption is customarily determined from batch slurry equilibrium assays. In this procedure, a small amount of medium is agitated in a solution having a known pesticide concentration. Agitation proceeds until apparent or real sorption equilibrium is achieved. The amount of chemical sorbed to the medium (C_s, μmol kg^{-1} or mg kg^{-1}) is determined by subtracting the pesticide concentration of the solution at the end of the agitation period (C_w, μmol L^{-1} or mg L^{-1}) from the pesticide concentration of the solution initially added to the media. This is done after accounting for the sample mass and the volume of solution used in the determination. Phase separation of C_s from C_w at the end of agitation is achieved by centrifugation or by passing the slurry through a 0.45-um pore size filter (*10*). Typically, several pesticide solution concentrations are examined to give C_ws that will span the likely concentration range of the pesticide in the media. When C_s is plotted as the function of C_w, the resulting relationship is referred to as a sorption isotherm.

Equilibrium sorption of a pesticide in soil is often quantified using the Freundlich equation. The simplest form of this equation is $C_s = K_f (C_w)^N$, where C_s and C_w are as previously defined, K_f ($\mu mol^{(1-N)} L^N kg^{-1}$ or $mg^{(1-N)} L^N kg^{-1}$) is the Freundlich sorption coefficient and N is an empirical constant that reflects the degree to which sorption is a function of equilibrium solution concentration. Although the Freundlich equation is considered an empirical model for sorption, N is often cited as being representative of a collection of adsorption site energies (*11-13*). The various site energies have been attributed to composition heterogeneity of the soil or media (*12*). Values of N that deviate from one are taken to indicate that different sorption phenomena are occurring at sites having different sorptive energy levels (*11*). Values of N between 0.75 and 0.95 frequently occur when a wide range of pesticide concentrations are examined and are thought to be due to declining site availability at higher pesticide solution concentrations (*13, 14*). Values of N above 1 occur less frequently in soil, and have been attributed to concentration-induced swelling of polymers that make up much of soil organic matter (*15*). The swelling of non-polar polymers presumably increases the partitioning of non-polar and slightly polar pesticides into these polymers (*15*). An alternative explanation for values of N above 1 is that a discrete organic phase of pesticide molecules develops around the sorbent (*16*) at high concentrations (ie., > 60 to 80% of the pesticide's solubility). The organic phase develops when pesticide molecules in solution bond to pesticide molecules already sorbed to the sorbent.

A special case of the Freundlich equation is when the sorbed concentration of a pesticide is directly proportional to the concentration of the pesticide in the solution (ie., N = 1). When this relationships is found to be valid by experimental means, or is assumed to be true, K_f is redefined as K_d, (L kg^{-1}), with K_d being referred to as the linear distribution coefficient, or the linear sorption coefficient. When N is equal to 1, all sites have equal adsorption energies. This implies that a single phenomenum is responsible for pesticide sorption to the medium. Linear sorption is typically observed when the range of the pesticide solution concentrations being examined is very narrow (*17*).

Extensive compilations of the pesticide sorption literature have demonstrated that much of the variation in sorption of slightly polar and non-polar pesticides in soils can be explained by the amount of organic matter present in the soil (*14, 18*). When pesticide sorption is assumed to be restricted to soil organic matter, the solid-liquid phase partition coefficient can be redefined as the organic carbon partition coefficient (K_{oc}), where K_{oc} is equal to K_d/f_{oc}, and f_{oc} is the fraction of organic carbon present in the soil. Oftentimes, K_f is used in place of K_d to calculate K_{oc}.

Thatch Sorption Measurements

Thatch resides above the soil surface and below sword verdure. It contains viable roots and stems as well as roots, stems and leaf sheaths in various states of decay. Thatch does not include the leaf blade or viable leaf sheath portion of the sword verdure. It does, however, include a modicum of soil deposited into the decaying plant material by wind, sediment deposition and insect activity. The addition of a large amount of soil to thatch usually gives rise to a thatch-like derivative call mat (19). The soil is typically introduced into thatch by intensive and prolonged earthworm activity, or by use management practices such as topdressing and/or hollow tine cultivation. Mat, when present, is generally considered to represent the lower boundary of decaying thatch organic matter.

Compared to soil and river sediment, relatively few measurements of thatch pesticide sorption exist in the literature. Thatch K_ds and K_fs published to date are presented in Table I. Four of the six studies summarized in Table I measured the sorption capacity of thatch only. The two remaining studies measured the lumped sorption capacity of the turf thatch and mat layer. The latter two studies also used a modified flow approach, rather that than the more traditional batch slurry procedure, to determine thatch pesticide sorption. The modified flow approach causes little disruption of the aggregates and organic matter compared to the batch slurry procedure. The former approach is thus thought to provide sorption estimates more closely approximating those found in the field (20). No direct comparison of the two procedures however has been published.

Two of the four studies that measured thatch pesticide sorption (versus thatch+mat) failed to identify the turf species that produced the thatch. Knowing such information may not be very important because, as previously mentioned, sorption of many pesticides is largely dependent on the organic carbon content of media. Raturi et al, (21) examined the sorption of triclopyr, 2,4-D and carbaryl to bentgrass thatch + mat and zoysiagrass thatch+ mat and found that the two media had similar pesticide sorptive capacities. The media however were comprised of different underlying soils and the thatch of the two turf species were not the same age. Additional studies free of the confounding effects of differing soil types, turf ages and management regimes would more clearly illuminate the importance of grass species on thatch pesticide sorption.

Spieszalski et al (22) determined the sorption of chlorpyrifos and fonofos to thatch by loss of the pesticide from the solution phase and by extracting and measuring the amount of pesticide sorbed to the thatch itself. Sorption determined by loss of the pesticide from the solution phase resulted in higher

Table I. Published Measured Thatch Sorption Coefficients

Ref.	Thatch Species	OC	Chemical	Solubility	Concentration Examined	K_f	N	K_d
		%		$mg\ L^{-1}$	$mg\ L^{-1}$	$\mu mol^{(1-N)}\ L^N\ kg^{-1}$		$L\ kg^{-1}$
(29)	Kentucky bluegrass	30	chloroneb	8[a]	0.10 to 1.6	163 (14)[b]	1.10 (0.09)	6453 (123)
			triadimefon	260[a]	0.16 to 2.3	91 (11)	0.86 (0.11)	4341 (647)[c]
			vinclozolin	1000[a]	0.14 to 2.2	432 (13)	1.74 (0.13)	592 (115)
(22)	not specified	34.8	chlorpyrifos	2	0.1 to 0.4	6166 / 3467[c]	0.99	398 (116)[c]
			fonofos	13	1 to 4	774 / 493[c]	0.83	
(23)	Kentucky bluegrass	39.8	acetanilide	5405[a]	~ 540 to 4324			3.53 (0.42)
			ethoprop	750[a]	~ 75 to 600			32.4 (2.40)
			1,2,4-tbc	49[a]	~ 4.9 to 39			167 (7.3)
			fenerimol	14[a]	~ 1.4 to 11			155 (13.3)
			phenanthrene	1[a]	~ 0.1 to 0.8			2520 (307)

Continued on next page.

Table I. Continued.

Ref.	Thatch Species	OC	Chemical	Solubility	Concentration Examined	K_f	N	K_d
		%		$mg\ L^{-1}$	$mg\ L^{-1}$	$\mu mol^{(1-N)}\ L^N kg^{-1}$		$L\ kg^{-1}$
(30)	not specified	35.6	atrazine	33[a]	4			41.7
			dicamba	4500	4			3.3
			2,4-D	900	0.8			32.4
			diazinon	60[a]	0.4			301
			chlorpyrifos	2[a]	4			592
(20)	zoysiagrass thatch+mat	14.2	dicamba	4500	1 to 20	0.82	1.26	
(21)	creeping bentgrass thatch+mat	7.7	2,4-D	900	1 to 100	3.0	0.87	
			triclopyr	2100000	0.1 to 100	2.6	0.86	
			carbaryl	40	1 to 300	43.4	1.22	

Zoysiagrass thatch+mat	7.3	2,4-D	900	1 to 100	2.9	1.04
		triclopyr	2100000	0.1 to 100	2.8	0.83
		carbaryl	40	1 to 300	46.4	1.12

[a] From values provided by authors
[b] Standard error of estimate
[c] Determined by extracting pesticide directly from thatch

194

K_ds and K_fs than did extracting the pesticide from the thatch itself. Higher sorption estimates obtained by simply measuring the change in pesticide concentration before and after the equilibration period was attributed to volatilization and degradation losses occurring during the equilibration period. The results of this study highlight the importance of minimizing sample vessel head space to reduce volatilization losses (23), and to quantify the formation of pesticide metabolites when evaluating sorption.

Sorption from Literature Values

In the absence of direct sorption measurements, consensus K_{oc} values, such as those typically found in pesticide property databases (24), are used to estimate pesticide sorption. Assumptions inherent in the use of K_{oc} to estimate pesticide sorption are that: 1) sorption is linear, 2) equilibrium is observed in the sorption-desorption process, 3) sorption and desorption isotherms are identical, 4) sorption is limited to the organic component of the media and 5) all organic carbon has the same sorption capacity per unit mass (25, 26). A natural question that arises is how well does the current thatch sorption data appear to satisfy these assumptions?

Sorption Linearity

Significant deviations from linearity have been reported for thatch pesticide sorption (Table I). Reported nonlinear isotherms for thatch are mostly limited to polar and ionic pesticides. This would appear to support the idea that non-polar and low polar pesticide sorption to thatch is the result of phase partitioning of the pesticide into partially decomposed non-polar plant material and any humidified soil organic matter that may be present in thatch. Detailed investigations of other types of high organic carbon content media (i.e., young peat soils) however, indicate linear sorption isotherms are usually the result of examining only a few C_ws, all of which are within two orders of concentration of the pesticide's solubility (27, 28). This was clearly demonstrated in a study that examined the sorption of 13 non polar compounds to a peat soil (15). In this study, at least 64 C_ws, spanning as much as six orders of concentration, were examined for six compounds. For each compound, running plots of the Freundlich N as a function of the fractional water solubility of the compound were constructed. The plots were constructed by fitting the Freundlich equation to data spanning two orders in concentration. The midpoint fractional water solubility of each isotherm was paired with the isotherm Freundlich N. In most cases, N was close to unity at very low fractional water solubilities. As the fractional water solubility of the compound increased, N declined until a minimum (ie., N = 0.55 to 0.85) was reached. Above the fractional water solubility associated with the lowest N value, N started to increase and approached or exceeded 1, as the fractional water solubility of the organic compound approached unity.

High water solubility pesticides having isotherms with N > 1 in thatch have been attributed to the divergent polarities of the sorbent and sorbate (21, 29). Conversely, nonlinear convex shaped isotherms reported for less polar carbaryl in thatch (21) are more likely the result of the high carbaryl concentrations used in the isotherm determinations. Divergent reports of the sorptive behavior of dicamba and 2,4-D in thatch (Table I) probably reflect the fact that pesticide sorption in one study (30) was evaluated at a single C_w. When one assumes *a priori* that sorption is linear, simple extension of the assumption sometimes results in the use a single C_w to determine K_d. The discussion in the preceding paragraph, however, highlights that K_d will usually depend on the C_w used in the evaluation. Farenhorst (31) for example, found the K_d of 2,4-D in a sandy loam soil decreased by a factor of ~ 4 when the single point C_w used to evaluate K_d was increased from ~ 1 mg L^{-1}, to ~ 33 mg L^{-1}.

Errors associated with using K_d when sorption is actually non-linear have been discussed by Hamaker and Thompson (18) and Rao and Davidson (14). It is generally recognized that when C_w is between 0.1 and 10 mg L^{-1}, sorption calculated using K_d will not deviate from that calculated using K_f by more than a factor of 3, when N is between 0.5 and 1.0. Errors of this magnitude have sometimes been considered tolerable for modeling agricultural applications (14). Recent model sensitivity evaluations, however, indicate that even relatively small deviations in N from unity can drastically alter the predicted pesticide leaching losses of some widely used models (1). For example, reducing N from 0.99 to 0.90 using PRZM reduced predicted pesticide leaching loss at 1 meter nearly 100 fold (ie., from 0.37 g ha^{-1} to 0.004 g ha^{-1}) for a model pesticide assigned a K_{oc} of 100 (1).

Assumed Equilibrium Conditions

Although K_d and K_f are frequently referred to as equilibrium distribution coefficients, it is generally acknowledged that the duration of most batch slurry assays is insufficient to fully consider all phases of a porous medium pesticide sorption reaction (28, 32). Very slow diffusion-driven reactions that occur in solids or soil aggregates for example, can take weeks to months to complete. From a practical standpoint, most soil batch sorption experiments are not extended beyond a day or two. This is because apparent (or near) sorption equilibrium is often observed for many pesticides by this time. At longer equilibrium times, pesticide biodegradation becomes a concern with many pesticides.

Thatch typically supports higher microbial populations than soil (*33, 34*). The larger pool of microorganisms present in thatch increases the probability that microorganisms capable of degrading specific pesticides will be present in this medium. The presence of such microorganisms can hasten pesticide transformations (*7*). Accordingly, errors in sorption caused by chemical transformations that occur during prolonged equilibration intervals may be more likely to occur in thatch than in soil.

Pesticide sorption that occurs over a day or two is assumed to be fully reversible (*28*). Many sorption investigations however, have reported that sorption distribution coefficients are typically lower than desorption distribution coefficients (*35*). A consequence of such behavior is delayed compound mobility within the media beyond what would be predicted by the batch sorption technique outlined earlier this chapter. Differing sorption and desorption isotherms have often been attributed to the methods used to obtain the two isotherms (*35*). In studies devoid of any obvious artifacts, isotherm non-singularity has been attributed to kinetically limited desorption, and/or to the entrapment of sorbed molecules within the organic or mineral components of soil (*25*). Desorption rate constants obtained from thatch+soil and soil-only column miscible displacement investigations (*20*), and from thatch and soil kinetics experiments (*36*) suggest that non-singularity may be more pronounced in thatch pesticide sorption/desorption isotherms than in isotherms obtained from soil underlying the thatch. Detailed kinetic investigations of the sorption and subsequent desorption of nonionic pesticides to thatch are needed to confirm this.

Sorption is Limited to the Organic Component of the Media

Pesticides containing weakly polar to strongly ionizable functional groups may react with polar or charged sites in organic matter and in mineral matter. Green and Karickoff (*26*) have suggested the contribution of mineral surfaces to the sorption of non-polar and weekly polar pesticides to soil becomes significant once the clay mineral to organic carbon content ratio of a soil exceeds 25 to 40. The high organic carbon contents of thatch and of thatch+mat (Table 1) effectively eliminate the possibility that mineral matter will be a significant sorbent of non-polar and weakly polar pesticides in these media.

Organic Carbon is Invariant in its Sorption Capacity

Several studies have reported that thatch organic carbon is a less effective sorbent of pesticides than is soil organic carbon (*20, 23, 29*). This is consistent with numerous studies that have shown the nature of organic matter affects the sorption of hydrophobic organic compounds (*37-39*). The lower sorptive capacity of thatch organic carbon compared to soil organic carbon has been attributed to the less decomposed state of thatch organic matter (*29*). Thatch is largely composed of cellulosic material and lignin (*40*). Cellulose is highly polar, while lignin is not. The low polarity of lignin makes this plant material the most likely site for the nonionic pesticide sorption to thatch. Higher pesticide sorption efficiencies seen in aged versus young straw, and in river sediment compared to soil, have been attributed to lower percentages of cellulose in the aged straw and river sediment organic matter (*41, 42*). This suggests the reduced sorption efficiency of thatch organic carbon, compared to soil organic carbon, is largely a consequence of the high cellulose content of thatch.

Dividing the thatch K_{oc} of a pesticide by the K_{oc} of the same pesticide in the underlying soil provides a means by which the sorption capacities of the organic carbon in two media can be compared. Figure 1 presents a compilation of the studies from Table I where the K_{oc} of thatch and the K_{oc} the soil underlying the thatch were determined for select pesticides. The thatch:soil K_{oc}, ratio for carbaryl from one study (*21*) has been omitted from Figure 1 as the value of this for this ratio was 1.68, and thus appeared to be anomalous when compared to the ratios calculated for all other comparisons. Although the data are limited in number, and include pesticides having nonlinear isotherms, it appears, on average, that thatch organic carbon has about 60% of the sorption capacity of soil organic carbon. More thatch and soil sorption investigations involving non-ionic pesticides are needed to better solidify the organic carbon relationships between these two media.

Use of Pesticide Properties to Estimate Sorption

Sorption of a chemical under consideration for use as a pesticide is often initially estimated from its chemical, physical and structural properties. The usual approach is to estimate the K_{oc} of the chemical from one or more of its physico-chemical properties. The resulting K_{oc} can then be used to estimate the K_d for a soil of interest. Numerous chemical and physical property relationships with K_{oc} have been established. Some of the relationships are specific to certain chemical classes (i.e., carbamates, triazines, ureas), while others have been shown to be applicable to wide variety of chemical classes (*25*).

The mostly widely used relationships to estimate K_{oc} are those that are based on water solubility (S), or the octanol-water partition coefficient (K_{ow}) of a pesticide (*43*). The K_{ow} of a chemical is the ratio of the chemical's concentration in 1–octanol to its concentration in distilled water. The K_{ow} is a measure of the hydrophobicity of the chemical. It can be determined experimentally (*44*), or by using the ClogP program if the structure of the molecule of interest is known (*45*; http://www.pirika.com/chem/TCPEE/LOGKOW/ourlogKow.htm). The

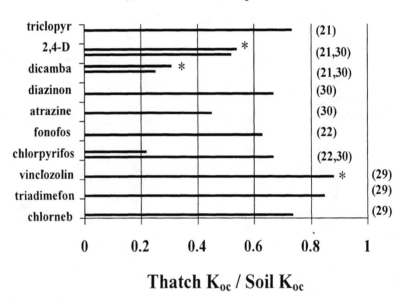

Figure 1. Thatch and soil K_{oc} ratios for studies where the liquid-solid phase partitioning coefficients were determined for both media

MedChem Master file also contains recommended K_{ow} values for over 10,000 chemicals in its LOGPSTAR database (43).

The use of pesticide physico-chemical properties to estimate thatch sorption has been sparingly investigated. Dell et al, (29) determined the K_{oc} of three pesticides (Table I) in thatch and soil and compared these values with those estimated by the relationship Log (K_{oc}) = 0.411*K_{ow}. The relationship was found to underestimate the K_{oc} of all three pesticides in both thatch and soil by a factor of 2.5 to 3.7. Lickfeldt and Branham (23) measured the sorption of 5 pesticides to thatch (Table I), and compared the experimentally determined K_{oc}s with the K_{oc}s obtained from 3 regression models. Two of the regression models estimated K_{oc} from S, while the third regression model estimated K_{oc} from S and the crystalline melting point of the pesticide. The latter regression model was the only model that did not consistently overestimate the K_{oc} of all five pesticides. This model, however, overpredicted the K_{oc} of three of the five pesticides by an average factor of 115. Near consistent overprediction of thatch K_{oc} by the three regression models was attributed to the fact that the regression models had been derived from soils sorption data rather than thatch sorption data.

Equations that estimate K_{oc} from the K_{ow} or S assume sorption is on the physico-chemical properties of the pesticide and the amount of organic carbon present in the medium. Recognition that sorption is also dependent on the source and degree of humification of soil organic matter has led to the development of relationships that consider the nature of organic matter as well (46-48). Comparison of the sorption of non-polar molecules in organic soils of varying age and origin has revealed that there is a strong negative correlation between the normalized sorption coefficient and the [oxygen (O)+nitrogen (N)/carbon (C)] content of the soils (46). The mass ratio of these three elements closely approximates the polar/non-polar balance of the soil (46). In principal, in young, decaying organic matter decreasing values of this ratio result from a decline in the fraction of cellulosic material present in organic matter. The lost of such material would be expected to be accompanied by a per unit weight increase in the sorption of non-polar molecules to the organic matter (46).

Xing et al (47) have referred to the (O+N)/C content of the organic phase of a soil as the "polarity index" of the soil. Widespread use of the polarity index to obtain soil specific K_{oc} estimates from K_{ow} and S relationships has not occurred because soil minerals contain large quantities of O and in some cases N. The presence of these sources of N and O make it difficult to easily determine the polarity index of the organic phase of mineral soils (47). Xing et al have emphasized that extraction of organic phase is never complete, and that the extraction process itself alters the polarity in regions where mineral and organic fractions of soils interact. This realization has resulted in most polarity index sorption investigations relying on the use of commercial biopolymers (i.e., lignin, specific humic acids, chitin, cellulose) and/or young organic soils (i.e., peat or muck) to examine relationships between aging soil organic matter and organic compound sorption (*46-49*).

From the perspective of future thatch sorptive investigations, the polarity index appears to offer a straightforward means by which the degree of humification of thatch can be assessed. The polarity index thus offers the possibility of serving as a unifying variable that can account for variances in the pesticide sorption capacity of different samples of thatch. Thatch polarity data could also be incorporated into existing K_{ow} and S relationships to improve pesticide sorption estimates for thatch.

The high organic matter content of thatch makes it ideally suited for polarity index sorption investigations. The polarity of this medium can be accessed in a manner similar to that already reported for organic soils (*46, 48*). Given the potential benefits that may be derived from knowledge of the polarity index, future thatch sorption studies should include measurements of the N, O and C contents of thatch.

References

1. Dubus, I.G.; Brown, C.D.; Beulke, S. *Pest. Manag. Sci.* **2003**, *59*, 962-982.
2. Durborow, T.E.; Barnes, N.L.; Cohen, S.Z.; Horst, G.L.; Smith, A.E. In *Fate and Management of Turfgrass Chemicals*; Clark, J.M.; Kenna, M.P., Eds.; ACS Symposium Series 743. American Chemical Society: Washington, DC, 2000, pp. 195-227.
3. Niemczyk, H.D.; Kruegar, H.R. *J. Econ. Entomol.* **1987**, *80*, 950-952.
4. Cisar, J.L.; Snyder, G.H. *Crop Sci.* **1996**, *36*, 1433-1438
5. Horst, G.L.; Shea, P.J.; Christains, N.E.; Miller, D.R.; Stuefer-Powell,; S.K. Starrett, *Crop Sci.* **1996**, *36*, 362-370.
6. Raturi, S.; Carroll, M. J.; Hill, *R.L. J. Environ. Qual.* **2003**, *32*, 215-223.
7. Roy, J.W.; Hall, J.C.; Parkin, G.W.; Wagner-Riddle, C.; Clegg, B.S. *J. Environ. Qual.* **2001**, *30*, 1360-1370.
8. Haith, D.A. *J. Environ. Qual.* **2001**, *30*, 1033-1039.
9. Gawlik, B.M.;Sotiriou, N. Feicht, E.A.; Schulte-Hostede, S.; Kettrup, A. *Chemosphere* **1997**, *34*, 2525-2551.
10. Designation D 46646-03. *Annual Book of ASTM standards* ASTM International. West Conshohocken, PA. 2004, Sec. 11, vol. 11.04, pp. 43-46.
11. Weber, W.J.; McGinley, P.M.; Katz, L.E. *Environ. Sci. Technol.* **1992**, *26*, 1955-1962.
12. Farrell, J.; Reinhard, M. *Environ. Sci. Technol.* **1994**, *28*, 53-62
13. Xing, B.; Pigmatello, J.I; Gigliotti, B. *Environ. Sci. Technol.* **1996**, *30*, 2432-2440.

14. Rao, P.S.C.; Davidson, J.M. In *Environmental Impact of Non-point Source Pollution*; Overcash, M.R.; Davidson, J.M., Eds.; Ann Arbor Science Publishers, Inc,; Ann Arbor, MI, 1980, pp. 23-65.

15. Xia, G.; Pignatello, J.J. *Environ. Sci. Technol.* **2001**, *35*, 84-94.

16. Hassett, J.J.; Banwart,W.L. In *Reactions and Movement of Organic Chemicals in Soils*; Sawhney, B.L.; Brown, K. Eds.; SSSA Special Publication No. 22. Soil Science Society of America, Madison, WI. 1989, pp. 31-44.

17. Weber , J.B.; Miller, C.T. In *Reactions and Movement of Organic Chemicals in Soils*; Sawhney, B.L.; Brown, K. Eds.; SSSA Special Publication No. 22. Soil Science Society of America, Madison, WI. 1989, pp. 305-334.

18. Hamaker, J.W.; Thompson, J.M. In *Organic Chemicals in the Soil Environment*. Hamaker, J.W.; Goring, C.A.I. Eds.; Marcel Dekker Inc, NY. pp.49-144.

19. Turgeon, A,J. *Turfgrass Management*. Pearson Prentice Hall, Upper Saddle River, NJ, 2005, pp.163-164.

20. Raturi, S.; Hill, R.L.; Carroll. M.J. *J. Soil and Sediment Contam.* **2000**, *10*, 227-247.

21. Raturi, S.; Islam, K.R.; Carroll, M.J.; Hill, R.L. *J. Environ. Sci. Health Part B*. **2005**, *40*, 697-710

22. Spieszalski, W.W.; Niemczyk, H.D.; Shetlar, D.J. *J. Environ. Sci. Health Part B*, **1994**, *29*, 1117-1136.

23. Lickfeldt, D.W.; Branham, B.E. *J. Environ. Qual.* **1995**, *24*, 980-985.

24. Wauchope, R.D.; Buttler, T.M.; Hornsby, A.G.; Augustijn-Beckers, P.W.M.; Burt, J.P. *Rev Environ. Comtam. Toxicol.* **1992**, *123*, 1-144

25. Doucette,W.J. *Environ. Toxicol. Chem.* **2003**, *22*, 1771-1788.

26. Green , R.E.; Karickhoff, S.W. In *Pesticides in the Soil Environment*; Cheng H. H. (Ed); Soil Science Society of American, Madison, WI, 1990, pp. 79-101.

27. Huang, W.; Young, T.T.; Schlautman, M.A.; Hong, Y.; Weber, W.J. *Environ. Sci. Technol.* **1997**, *31*, 1703-1710.

28. Wauchope, R.D.; Yeh., S.; Linders, J.; Kloskowski, R.; Tanaka, K.; Rubin, B.; Katayama, A.; Kordel, W.; Gerstl, Z.; Lane, M.; Unworth, J.B. *Pest Manag. Sci.* **2002**, 419-445.

29. Dell, C.J.; Throssell, C.S.; Bischoff, M.; Turco, R.F. Turco. *J. Environ. Qual.* **1994**, *23*, 92-96.

30. Baskaran, S.; Kookana, R.S.; Naidu, R. *Intern. Turf Soc. Res. J.* **1997**, *8*, 151-165.

31. Farenhorst, A. *Soil Sci. Soc .Am. J.* **2006**, *70*, 1005-1012.

32. Smith, M. C.; Shaw, D.R.; Massey, J.H.; Botette, M.; Kingery, W. *J. Environ. Qual.* **2003**, *32*, 1393-1404.

33. Mancino, C.F.; Barakat, M.; Maricic, A. *HortScience.* **1993**, *28*, 189-191.

34. Raturi, S.; Islam, K.R.; Carroll, M.J.; Hill, R.L. *Commun. Soil Sci. Plant Anal.* **2004**, *35*, 2161-2176.
35. Pignatello, J,J. In *Reactions and Movement of Organic Chemicals in Soils*; Sawhney, B.L.; Brown, K. Eds.; SSSA Special Publication No. 22. Soil Science Society of America, Madison, WI. 1989, pp. 45-80.
36. Raturi, S., Carroll, M.J., Hill, R.L., Pfeil, E. and Herner. A.E., *Intern. Turf. Soc. Res. J.* **1997**, *8*, 187-196.
37. Garbarini, D.R.; Lion, L.W. *Environ. Sci. Technol.* **1986**, *20*, 1263-1269.
38. Grathwohl, P. *Environ. Sci. Technol.* **1990**, *24*, 1687-1693.
39. Kookana R.S; Ahmad, R. In *Proceedings 17th World Soils Congress*, Bangkok, Thailand. 2002, Paper no. 1933,, pp. 1-10.41.
40. Ledeboer, F.B.; Skogley, C.R. *Agron. J.* **1969**, *59*, 320-329.
41. Doa, T. *J. Environ. Qual.* **1991**, *20*, 203-208.
42. Gerstl, Z.; Kliger, L., *J. Environ. Sci. Health Part B,* **1990**, *25*, 729-741.
43. Sabljic, A.; Gusten, H.; Verhaar, H.; Hermens, J. *Chemosphere,* **1995**, 4489-4514.
44. Shey, P.J. *Weed Sci.* **1989**, *3*, 190-197.
45. Leo, A.J. *Chem. Rev.* **1993**, 1281-1306.
46. Rutherford, D.W.; Chiou, C.T.; Kile, D.E. *Environ. Sci. Technol.* **1992**, *26*, 336-340.
47. Xing, B; McGill, W.B.; Dudas, M.J. *Environ. Sci. Technol.* **1994**, *28*, 1929-1933.
48. Chen, Z. Xing, B; McGill, W.B.; Dudas, M.J. *Can. J. Soil Sci.* **1996**, *76*, 513-523.
49. Torrents, A.; Jayasundera, S.; Schmidt, W.J. *J. Agri. Food Chem.* **1997**, *45*, 3320-3322.

Chapter 12

Regional Analyses of Pesticide Runoff from Turf

Douglas A. Haith[1], Matthew W. Duffany[2], and Antoni Magri[3]

[1]Biological and Environmental Engineering, Cornell University,
Riley-Robb Hall, Ithaca, NY 14853
[2]New York State Department of Environmental Conservation,
317 Washington Street, Watertown, NY 13601–3787
[3]ESRI, 380 New York Street, Redlands, CA 92373–8100

Pesticide runoff loads from turf can vary dramatically with
chemical properties and application regime, geographic
location, irrigation rates and turf surface. Given the limited
availability of field data, it is difficult to realistically consider
the range of these variations in exposure assessments. The
TurfPQ pesticide runoff model was combined with several
other models and data bases to provide a general framework
for efficient estimation of turf pesticide runoff loads on both a
yearly and daily basis. The process was used to investigate
differences in MCPP, fenarimol, iprodione and carbaryl runoff
from fairways at four U.S. locations with widely differing
climatic regions. Factors which accounted for the observed
differences included pesticide properties and application
amounts, irrigation applications and growing season runoff.
The simulations indicated that runoff loads of a particular
pesticide could vary by as much as an order of magnitude
among the locations.

One of the significant difficulties in managing the environmental impacts of turf pesticide runoff is the immense variability in transport and fate characteristics. One pesticide may be easily washed from grass surfaces by small amounts of runoff while another resists movement, even with extreme storms. Some chemicals persist in the turf and soil for months while others are degraded within days or even hours. These variations are further compounded by differences in weather patterns between geographic locations. As a result, a program for controlling the runoff of one pesticide at one site is not likely to be adequate for another chemical and site.

A classic approach for elucidating such differences is through controlled field experiments. Given the large number of available turf pesticides and the many different weather regimes seen in an area as large as the continental U.S., this approach has limited practicality. Fortunately, many of its features can be duplicated in simulation experiments. Mathematical models are used to describe weather and runoff, and the effects of a variety of site conditions and management options can be efficiently evaluated. Nevertheless, simulation experiments of pesticide runoff are challenging. Available models often require many input parameters whose values are difficult to estimate. Information on the rates and timing of pesticide applications for a particular location may be particularly difficult to obtain.

The research described herein had two objectives. The first was to develop a general protocol for simulation studies of pesticide runoff from turf. The protocol is built around the TurfPQ pesticide runoff model *(1,2)*, and USCLIMATE weather generator *(3)*, but the methods should be applicable to other models as well. The second objective was to demonstrate the protocol through a simulation experiment designed to study the regional differences in runoff of several pesticides applied to fairways.

Simulation Protocol

A simulation protocol consists of the design of the simulation experiment's scenario (pesticide selection, site description, length of simulation run), the specification of appropriate models, estimation of input parameters, and selection of methods for summarizing and interpreting results.

Scenario

Four pesticides were simulated: the herbicide MCPP (2-(2-Methyl-4-chlorophenoxy) propionic acid), two fungicides, fenarimol (α-(2-Chlorophenyl)-α-(4-chlorophenyl)-5-pyridinemethanol) and iprodione (3-(3,5-Dichlorophenyl)-

N-(1-methylethyl)-2,4-dioxo-1-Imidazolidinecarboxamide), and the insecticide carbaryl (1-Naphthyl-N-methylcarbamate). The sites are identical, hypothetical golf fairways in Atlanta, GA; Fresno, CA; Madison, WI; and Olympia, WA. Weather characteristics for these sites are given in Table I. Temperatures and precipitation are 1971-2000 means *(4)*. Growing seasons are based on median freeze/frost dates *(5)*.

Table I. Weather Characteristics of Simulation Sites

Location	Annual Temperature (°C)	Annual Precipitation (mm)	Growing Season
Atlanta, GA	16	1290	Apr-Oct
Fresno, CA	17	270	Mar-Nov
Madison, WI	7	785	May-Sep
Olympia, WA	10	1285	May-Oct

It can be seen from Table I that the four sites have substantially different weather characteristics. Atlanta and Fresno both have warm climates, but Fresno is much drier and would require significant irrigation to maintain turf surfaces. Madison and Olympia are cooler and have shorter growing seasons than the other cities. Although Olympia's annual precipitation is comparable to Atlanta's, it is differently distributed. Atlanta precipitation is relatively uniformly distributed throughout the year, but Olympia has little growing season moisture. Unlike field experiments, simulations can be of any duration. It is typically as easy to make 500-year runs as 5-year ones. In general, runs should be long enough to provide reliable estimates of the phenomena of interest. In the current study, regional differences were evaluated by comparison of annual and monthly means and 1 in 10-year extreme events, and these variables could be reasonably estimated from 100 years of daily results. This does not imply that the experiments modeled 100 years of fairway operations. Rather, the 100-year run should be interpreted as producing 100 different estimates of one-year of pesticide runoff.

Simulation Models

The TurfPQ model was used in this study to simulate pesticide runoff. The model computes water and chemical mass balances on a one-day time step. Runoff volume is determined through a modified curve number equation. Pesticide in turf foliage and thatch is partitioned into adsorbed and dissolved components which are assumed to be decayed in a first order biodegradation process. In addition to decay, dissolved pesticide is removed from the system by runoff or leaching into the soil. Volatilization is neglected. In addition to daily precipitation and temperatures and pesticide application rates, the model requires four input parameters – biodegradation half-life, organic carbon partition coefficient, runoff curve number, and organic carbon content of the turf. In a validation study of 52 runoff events in four states involving 6 pesticides, TurfPQ explained 65% of the observed variation in pesticide runoff. Mean predicted pesticide runoff was 2.9% of application, compared to a mean observation of 2.1% (1,2).

The USCLIMATE software package, which was used to generate daily weather data for the TurfPQ model, produces daily precipitation, minimum and maximum air temperatures and a solar radiation record for arbitrary user-specified locations in the continental U.S. Precipitation is based on a Markov chain of occurrence (wet/dry days) and a mixed exponential distribution for precipitation amount. Temperatures are described by an autocorrelation model conditioned on wet or dry days. The generated weather data are processed in several ways to produce the daily records of precipitation and temperatures required by TurfPQ. Solar radiation data are discarded and the software's March to April sequences are converted to January to December. Daily temperatures are obtained by averaging the minimum and maximum temperatures.

Input Data

Weather

Depending on the nature of the site, the weather records may be further modified to reflect the addition of irrigation. This would generally be the case for golf course turfs. In this study, irrigation was based on comparison of 3-day cumulative precipitation and potential evapotranspiration during the growing season. Whenever the 3-day precipitation is exceeded by 3-day potential evapotranspiration as computed by the Hamon equation (6), irrigation is added to make up the deficit. This produces a new weather record in which precipitation entries for any day are replaced by precipitation plus irrigation.

Turf Properties

Turf properties required for the simulations are runoff curve number for average antecedent moisture conditions (CN2) and the organic carbon content of the grass and thatch. Both of these parameters depend on grass height and thatch thickness, which were assumed to be 11 and 8 mm, respectively, as in Haith and Rossi *(7)*. Using the procedures given in Haith *(1)*, these values produce a curve number of 67 and organic carbon content of 10,200 kg/ha. The curve number selection also assumes a hydrologic group C (relatively poor drained) soil.

Pesticide Characteristics

The two pesticide properties required by TurfPQ, bio-degradation half-life and partition coefficient, are relatively easily obtained. The partition coefficient is computed from turf organic carbon content and K_{oc}, the organic carbon partition coefficient. Half-lives and K_{oc} values are available from general databases *(8,9,10)*. Application amounts and timing are also required for the simulations, and these can be quite difficult to obtain. Although application rates are specified by labels *(11)*, a wide range is often given, corresponding to use against different pests. Because it is likely that the chemicals will typically be used against a variety of pests, the median or mid-range label value, converted to g/ha of active ingredients, was used in the simulations.

Timing, or frequency of applications, is less straightforward. Publicly available application records are very rare, and we know of no general databases. In the absence of other information, we based simulation applications on label suggestions of prophylactic applications at regular intervals to control multiple pests. These applications will almost certainly be more frequent than those used by many turf managers, particularly those following integrated pest management programs. The major determinants were pesticide type (herbicide, fungicide, insecticide), growing season, as shown in Table I, and application intervals and annual or seasonal limits specified by the labels. Generally, longer growing seasons result in more applications of a pesticide, unless limited by label.

Herbicides are divided into pre-emergent and post-emergent. The former is applied as a single application on the first day of the growing season. Post-emergent herbicides such as MCPP are assumed to be applied in the middle of each of the first two months of the growing season and once in the last or next to last month of growing season if allowed by the label.

Fungicide applications were based on preventative control of diseases such as dollar spot, summer patch, brown patch, and leaf spot. Applications were generally started in the middle of the second growing month, and if permitted by label, continued every 15 days through the middle of the next to last growing month. Otherwise, label limits applied, as was the case with fenarimol, which was applied every 30 days.

As with fungicides, repetitive preventive applications are assumed for insecticides, which are used to control a range of pests (grubs, chinch bugs, cutworms, webworms, billbugs) which occur mainly in late Spring and Summer. For insecticides such as carbaryl, this suggests a mid-month application starting in the second growing season month and continuing through September. Pesticide properties, rates and application frequencies for the four simulated chemicals are given in Tables II and III.

Table II. Pesticide Properties and Application Rates

Pesticide	Rate per Application (g/ha)	Half-Life (d)	K_{oc} (cm³/g)
MCPP	860	10	20
Fenarimol	760	840	760
Iprodione	4580	50	670
Carbaryl	8000	17	290

Application rates and frequencies differ markedly for these four chemicals. For example, the total annual pesticide application for the Atlanta site ranges from 2580 g/ha for MCPP to 40,000 g/ha for carbaryl. Application frequency is lowest for Madison because of its short growing season. This produces much lower inputs of the 2 fungicides than seen at the other sites. The large number of fungicide applications for Fresno may seem inconsistent with its dry climate, which would not typically favor plant diseases. However, the regular irrigation inputs needed to maintain Fresno fairways produce the warm, humid conditions required for disease development.

The pesticides differ markedly in their persistence and adsorption characteristics (half-lives and K_{oc}). MCPP is an ephemeral chemical that is only weakly adsorbed, and unlikely to remain long in the turf. Carbaryl is similarly short-lived, but more strongly adsorbed and thus less readily leached. Both fungicides are relatively strongly adsorbed, and fenarimol is very long-lived.

Table III. Pesticide Application Frequency

| Pesticide | Number of Applications | | | |
	Atlanta	Fresno	Madison	Olympia
MCPP	3	3	3	3
Fenarimol	5	7	3	4
Iprodione	6	6	3	5
Carbaryl	5	6	4	4

Organization of Results

Each simulation experiment produces 100 years of daily estimates of water volumes and pesticide mass loads in fairway runoff. The information was summarized by annual and monthly means and by the annual maximum daily load (AMDL) of pesticide runoff. The AMDL is the largest one-day runoff load produced in a year. The 100 values of AMDLs are then used to assign return periods to these extreme event. Thus the 1 in 10 year AMDL would be expected to be exceeded on the average of once in 10 years, or 10 times in 100 years.

Simulation Results

Annual Water Balances

Mean annual water inputs and runoff from the 100-year simulations are given in Table IV. Overall, regional differences in weather and hydrology for these sites are rather substantial. Runoff was minimal for Fresno because most water input was from the regular addition of moderate irrigation amounts rather than large precipitation events. Runoff was 3-4% of total water inputs at the other sites, and 40-50% of the runoff occurred during the growing seasons at Atlanta and Madison. Although Olympia had significant annual runoff, very little occurred during the growing season, when pesticides were being applied.

Table IV. Mean Annual Fairways Water Inputs and Runoff

Location	Precipitation	Irrigation	Total	Year Runoff	Growing Season Runoff
			---mm---		
Atlanta	1281	435	1716	77	34
Fresno	272	771	1043	2	<1
Madison	789	307	1096	32	16
Olympia	1304	330	1634	65	4

Annual Pesticide Runoff

The mean annual pesticide mass loads in runoff from the 100-year simulations are shown in Figure 1. To a considerable extent, these results reflect the differences in application and runoff water amounts. Chemicals such as

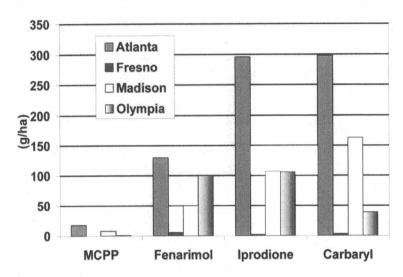

Figure 1. Mean annual runoff of four pesticides at four fairways sites.
(See page 3 of color inserts.)

iprodione and carbaryl, which are applied in greatest amount, are most likely to be seen in runoff. Similarly, sites such as Fresno, which have very little runoff, have correspondingly small amounts of pesticide loss. However, regional differences are not always clear-cut. Madison and Olympia produced almost equal amounts of iprodione runoff, but differed greatly in fenarimol and carbaryl runoff. Fenarimol runoff was 50% higher in Olympia but carbaryl runoff was 4 times higher in Madison. These apparently inconsistent results reflect the chemicals' properties and occurrences of runoff water. Carbaryl is short-lived and most likely to be lost during the growing season in which it is applied. Growing season runoff is much higher in Madison, so there will be greater opportunities for loss. Conversely, the persistence and strong adsorption of fenarimol means that it is likely to remain available for runoff following the growing season, when Olympia experiences much greater runoff than Madison.

Comparisons of mean annual pesticide runoff look rather different when the mass loads are normalized with respect to total annual application, as shown in Figure 2. Three of the pesticides, MCPP, iprodione and carbaryl, show similar tendencies for loss in runoff, but these losses are much smaller than those seen for fenarimol. It is apparent that the relatively larger mass runoff loads of iprodione and carbaryl that were seen in Figure 1 were more a result of the larger applications of the chemicals than their inherent propensities for loss.

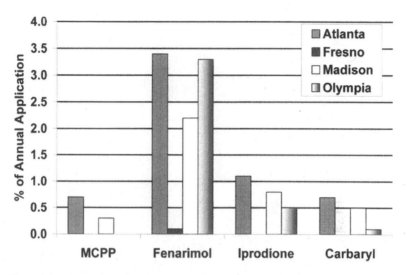

Figure 2. Mean annual fairway runoff pesticides expressed as a percentage of annual application. (See page 3 of color inserts.)

The regional differences in the normalized runoff results are consistent with those observed for mass loads. Atlanta still produces the greatest runoff losses, but the differences with other locations are less striking. Atlanta mass loads were more than twice as large as Madison's for all chemicals, but when expressed as a percentage of annual applications, the differences are much less, particularly for iprodione and carbaryl. Averaged over all chemicals, mean annual losses by location are Atlanta – 1.5%, Fresno – <0.1%, Madison and Olympia – 1.0%. Mean chemical losses, averaged over location, are MCPP – 0.3%, fenarimol – 2.3%, iprodione – 0.6% and carbaryl – 0.3%.

Pesticide Runoff in Extreme Events

Although water quality impacts are often measured in terms of mean annual loads such as those shown in Figure 1, these indicators may be of limited value for turf pesticide runoff. Most turf systems, including the fairways modeled in these simulation experiments, produce water runoff infrequently, and significant pesticide runoff is produced only when one of these events coincides with a high level of available chemical in the turf foliage and thatch. There are typically few such occurrences in any year, but it is these short-term phenomena that are responsible for any impact that pesticide runoff will have on surface

212

receiving waters. Mean annual loads are useful indicators of the relative likelihood of pesticide runoff, but they are imperfect measures of impact.

Figure 3 shows the 1 in 10 year, 1-day pesticide runoff event modeled for the four chemicals and sites. This is the event that is likely to occur, on average, once every 10 years. These results look very different than the mean annual values shown in Figure 2. Fenarimol remains the chemical with highest percentage loss at three locations, but MCPP runoff in Atlanta exceeds that of the other chemicals. Atlanta no longer sees the largest losses for all pesticides. Madison produces the greatest percentage losses of fenarimol and iprodione, even though runoff water volume is much lower than for Atlanta and Olympia.

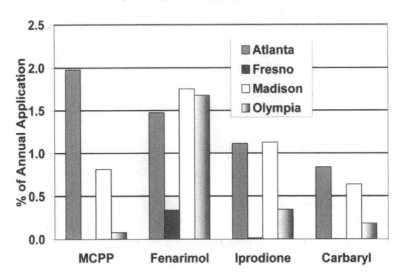

Figure 3. One in ten year pesticide runoff events.
(See page 4 of color inserts.)

The regional and chemical variations seen in the extreme event results in Figure 3 are not readily explained by differences in pesticide properties or mean weather conditions. Rather, they are influenced by the nature of the precipitation and snowmelt events at the site, and these in turn are determined by the probabilistic structure of the weather. Are large storms more likely to occur shortly after pesticide application? Is a site characterized more by large numbers of small storms rather than by rare large events? The effects of these weather characteristics on pesticide runoff and the resulting water quality impacts are essentially unpredictable, but we can observe them through longterm simulations such as these.

Conclusions

Although field and simulation experiments have comparable objectives and designs there is at least one profound difference. Field experiments measure reality and simulation experiments estimate reality. Granted, measurements can be error-prone and misconstrued, and simulation models are often tested through field experiments, but still, one is real and the other is not.

So what are the conclusions that can be drawn from these simulation experiments? First, it would be foolhardy to greatly trust the absolute values of the results. Because of the intensive pesticide application frequencies and poorly drained soils assumed in the simulations, the resulting pesticide runoff loads are likely to be larger than those seen on many sites. Further, although the TurfPQ model is relatively accurate on average, its estimates can differ substantially from field measurements for any one site or chemical. For example, we know that its estimates of runoff of strongly adsorbed pesticides are often too high (1,2). However, the model does better at estimating differences among chemicals and sites, and that is the critical attribute for the present study.

The major conclusion of the simulation experiment is that substantial differences exist in turf pesticide runoff among sites and chemicals. Stated another way, conclusions about runoff of a pesticide at one location cannot safely be extrapolated to another pesticide or location. This is true whether we are talking of mean mass loads, percentage losses or extreme events. Some of the differences can be explained by chemical properties and annual weather and hydrology conditions, but others, including those seen in extreme pesticide runoff events, are unpredictable, and only become apparent with long-term observations.

This is not a happy conclusion for the turf manager or government regulator wishing to adopt general guidelines for environmentally safe pesticide use. It argues that each combination of chemical, application regime and site is unique. Field experiments for each of these situations would be immensely impractical, but simulation experiments are much less forbidding. The simulation protocol described in this paper is a straightforward intuitive process that is accessible to anyone with computer skills and basic knowledge of pesticides and turf. The process could be made even more accessible if captured in a web-based interactive system. It may be time to admit that generalizations regarding environmental impacts of turf pesticides are neither desirable nor necessary. The methods and data are available for quantitative analyses of each unique situation.

References

1. Haith, D. A. *J. Environ. Qual.* **2001**, *30*, 1033-1039.
2. Haith, D. A. *J. Environ. Qual.* **2002**, *31*, 701-701.
3. Hanson, C. L.; Cumming, K. A.; Woolhiser; D. A., Richardson; C. W. *Microcomputer Program for Daily Weather Simulation in the Contiguous United States;* ARS-114. UDSA-ARS: Washington, D.C. 1994.
4. National Oceanographic and Atmospheric Administration. *Comparative Climatic Data*. URL http://ols.nndc.noaa.gov/plolstore/. 2005.
5. Koss, W. K., Owenby, J. R., Steurer, P. M., Ezell, D. S. *Freeze/Frost Data;* Climatology of the U.S. No. 20, Supplement No. 1. National Climatic Data Center: Ashville, NC. 1988.
6. Hamon, W. R.. *Proc. Am. Soc. Civ. Eng., J. Hyd. Div.* **1961**, *87*(HY3), 107-120.
7. Haith, D. A.; Rossi, F.S. *J. Environ. Qual.* **2003**, *32*, 447-455.
8. Agricultural Research Service. *The ARS pesticide properties database*. URL http://www.arsusda.gov/acsl/services/ppdb/. 2005.
9. *The Pesticide Manual;* Tomlin, C., Ed. British Crop Protection Council: Farnham, U.K. 2000.
10. Natural Resources Conservation Service. *Pesticide properties database*. URL http://www.wcc.nrcs.usda.gov/pestmgt/. 2005
11. Vance Communication. *Greenbook Database of Labels, MSDS, and Supplemental Labels,* URL http://www.greenbook.net/. 2005

Chapter 13

Mobility and Dissipation of Clopyralid Herbicide in Turfgrass Field Lysimeters

David W. Roberts[1], Ian J. van Wesenbeeck[2], James M. Breuninger[2], and Thomas Petrecz[3]

[1]Pioneer Hi-Bred International, 7250 NW 62nd Avenue, Johnston, IA 50131–0552
[2]Regulatory Laboratories, Dow AgroSciences LLC, 9330 Zionsville Road, Indianapolis, IN 46268
[3]Penn E&R Environmental and Remediation, Inc., Hatfield, PA 19440

The mobility and dissipation of clopyralid herbicide was studied in undisturbed soil monoliths (lysimeters) on Long Island, NY, June through October, 2000. Clopyralid was applied at a rate of 280 g acid equivalent (ae) per hectare to lysimeters containing undisturbed sand, considered highly vulnerable to leaching, with turfgrass cover. Bromide (Br^-) was applied to track water movement. Natural rainfall was supplemented weekly with sprinkler irrigation to achieve at least 130% of normal precipitation. Downward movement of clopyralid was limited to the upper 15-cm soil profile, dissipating rapidly in soil with a first-order field half-life of 9 days. For the entire 4-month study period, clopyralid was undetected (detection limit 0.4 ppb) in leachate even after rainfall plus irrigation totaled 552 mm or 173% of normal. Cumulative leachate volume totaled 310 mm, or 56% of on-site rainfall plus irrigation, and bromide tracer breakthrough was well established. Although some physico-chemical properties of clopyralid are characteristic of pesticides having a potential to leach, this study demonstrated that leaching potential in the field, under turfgrass use conditions, is extremely low, being offset by plant interception, rapid soil degradation and/or increased soil sorption with increased residence time.

Clopyralid (chemical name: 3,6-dichloropyridine-2-carboxylic acid) is a systemic, auxin type of growth regulator herbicide used as a selective herbicide for pre-emergence, pre-plant or post-emergence control of annual and perennial broadleaf weeds in small grains, corn, sugar beets, fallow cropland, rangeland, pastures, and non-cropland (e.g., rights-of-way, non-residential turf and ornamentals). Commercial formulations contain clopyralid in the free acid form, or the amine-salt formulation, which dissociates rapidly (with a pK_a value of 2.0) to the clopyralid free acid in environmental pHs greater than 2.0. Clopyralid is currently marketed by Dow AgroSciences in the U.S. as Lontrel™ Turf and Ornamental Herbicide and as Confront™ for non-residential ornamental and turf use. Other formulations are Transline™ Specialty Herbicide for rights-of-way use patterns, and Stinger™ Herbicide for crop use patterns. Maximum annual broadcast use rate of clopyralid is 3-4 oz (ae) acre^{-1} or 200-300 g (ae) hectare^{-1}.

Selected physico-chemical properties, such as water solubility greater than 30 ug/L (Table I) (1), are characteristic of pesticides having a potential to leach (2). However, key properties governing leaching potential are sorptivity (K_d or K_{oc}) and decay rate (aerobic soil metabolism and field dissipation half-life). The very low sorption potential of clopyralid (K_{oc}: 0.4-12.9 ml/g) is significantly offset by an average half-life of 22 days observed among 20 field sites. Also, when aged in soil for 30 days, a K_{oc} of 30 ml/g was observed, suggesting clopyralid is more tightly sorbed with time in soil. Therefore, in most environments, the greatest leaching potential is restricted to the initial few days after application. These conclusions are supported by results of field dissipation studies, where downward movement was restricted to 45 cm of the soil profile (Table I).

Undisturbed soil monoliths (lysimeters) have been used extensively in the U.S. and Europe as a viable alternative to field-scale ground water studies in evaluating the mobility of compounds under actual field conditions (3). In lysimeter studies conducted in Europe using ^{14}C-labeled clopyralid applied post emergence to agricultural crops at labeled use rates, downward movement was limited to 50 cm after 2-3 years, and average annual concentrations in leachate were 0.001-0.055 ppb (4, 5, 6, 7, 8, 9, 10, 11). These lysimeter studies concluded that in worst-case, sandy soils under typical crop use patterns, clopyralid degradation was sufficient to limit the risk of groundwater contamination to levels well below the European Commission (EC) drinking water limit of 0.1 μg L^{-1}, or for clopyralid in leachate, to levels below 0.1% of the amount applied.

In the U.S. and Canada, no Health Advisory Limit (HAL) or Canadian Water Quality Guideline has been set for clopyralid, reflecting the low human health and ecological concerns. As a result, the 0.1 μg L^{-1} EC drinking water

---

TM Registered Trademark of Dow AgroSciences LLC.

Table I. Properties of Clopyralid

Property	Value
Common Name:	clopyralid
Chemical Name (IUPAC):	3,6-dichloropyridine-2-carboxylic acid
Structural Formula:	
Empirical Formula:	$C_6H_3C_{12}NO_2$
Molecular Weight:	$192.0 \ g \ mol^{-1}$
CAS Number:	001702-17-6
D.I. Water Solubility (20 °C):	$S_w = 7.85 \ g/L$
Water Solubility (20 °C):	$S_w = 118 \ g/L$ (at pH 5) to 157 g/L (at pH 9)
Dissociation Constant:	$pK_a = 2.0$ (25 °C)
Vapor Pressure (25 °C):	$V_p = 1.02 \times 10^{-5}$ mm Hg; 1.36×10^{-6} kPa
Henry's Law Constant (20 °C):	$K_H = 3.28 \times 10^{-10}$ atm m^3/mol
Octanol/Water Partition Coefficient:	$\log K_{ow} = -1.81$ (at pH 5); -2.63 (at pH 7) -2.55 (at pH 9)
Hydrolysis (sterile water):	Half-life > 30 days at 25 °C (pH 5-9)
Photolysis (soil and aqueous):	Stable
Plant Metabolism:	not metabolized in plants
Soil Distribution (Sorption) Coefficient:	$K_d = 0.01 - 0.09$ mL/g
Soil Organic Carbon Sorption Coefficient:	$K_{oc} = 0.4 - 12.9$ mL/g (24-hr test) 29.8 mL/g (after 30 days)
Aerobic Soil Metabolism:	Half-life = 2 - 94 days No metabolites other than CO_2. Degradation faster in warm, moist soils and lower application rates. Half-life = 7 days (at 0.0025 ppm) to 435 days (at 2.5 ppm) in sandy loam
Field Soil Dissipation:	Average Field Half-life = 22 days (7-66 days at 20 sites); maximum depth of detection 45 cm

NOTE: Data in this table are from reference 1.

limit does not apply for U.S. or Canada regulatory purposes. In the U.S., individual states have the authority to set there own drinking water limits. For example, New York state has set a drinking water limit of 50 ppb. Regulators there used an EPA-approved model, LeachP, to predict a soil water concentration (below a 90 cm soil profile) of 25 ppb for clopyralid (*12*). However, models such as LeachP and PRZM were intended for use as a screening-level assessment of leaching potential, and do not account for mixing, dilution or raw water treatment for distribution as drinking water. Also, the models do not account for increased soil sorption with time. For these reasons, a higher-tiered investigation of leachability through the use of field lysimetry was the natural progression for an accurate assessment of real-world impact.

The purpose of this study was to gather information on the dissipation and movement of clopyralid under field conditions representative of non-crop (grassland or recreational) turfgrass use patterns.

Materials and Methods

This report is based on an unpublished report of Dow AgroSciences (*13*), and was conducted in compliance with Good Laboratory Practice Standards (GLPS) as defined by 1) United States Environmental Protection Agency, Title 40 Code of Federal Regulations Part 160, Federal Register, August 17, 1989, and 2) Organisation for Economic Co-Operation and Development, ISBN 92-64-12367-9, Paris 1982.

Study Site

One test site was located on the golf coarse property of Maidstone Club in East Hampton, Suffolk County, NY, located at the northeast end of Long Island (latitude N 40° 57' 09", longitude W 72° 10' 45"). The test site was on a mature stand of tall fescue (*Festuca arundinacea* Schreb.) maintained as a golf course rough. The soil at the test site was of the Plymouth soil series (Mesic, coated Typic Quartzipsamments). A typical Plymouth series profile contains loamy sand to 70 cm, and gravelly coarse sand to 147 cm, and covers 19,076 acres on Long Island, the third largest soil series coverage for Long Island. This soil is excessively drained, and permeability is rapid in the solum and very rapid in the underlying substratum. Soil texture specific to the test site was classified as sand throughout the 90 cm profile. Organic matter generally decreased from 2.6% at the surface to 0.4% at 90 cm. Soil pH ranged from 5.4 to 5.6 through 90 cm. Bulk density (disturbed soil) ranged from 1.41 g cm^{-3} in the surface 15-cm depth segment to 1.58 g cm^{-3} in the 75-90-cm segment. Average 30-year normal annual rainfall in this location is 1168 mm (46 in.), and the normal annual temperature is 19 °C (51°F).

Lysimeter Design, Installation and Management

The test site consisted of nine 30 cm id x 100 cm length lysimeters, eight of which were designated for test and tracer substance treatment and subsequent sampling of leachate or soil. One was designated as an untreated control. Lysimeters were installed using a hollow-stem auger method generally described by Persson and Bergstrom (*14*) and fitted with a leachate collection assembly to produce a lysimeter. Each lysimeter was returned to its source location, and back-filled around the lysimeter using soil from the source location. The surface of the turfgrass within the lysimeter was equal to the surface outside the lysimeter. In one lysimeter (the untreated control), volumetric soil water content was measured using a time domain reflectometry (TDR) system. Prior to returning the lysimeter to its source location, three Campbell (Campbell Scientific, Logan, Utah) CS-615 sensors were inserted through holes drilled in the side of the lysimeter at the 15, 45 and 75 cm soil depth. The soil moisture sensors were connected to a Campbell 21XL® microprocessor (data logger), and measurements were taken every 30 minutes. The lysimeters were allowed to equilibrate for one month prior to test substance application. During the equilibration interval, the lysimeters were periodically inspected to determine uniformity of leachate volumes. Because none of the lysimeters exhibited hydraulic anomalies, no lysimeter was rejected from the study. Turfgrass sod consisting of tall fescue was collected from the surrounding area and established in each lysimeter 2 weeks prior to application. The grass in each lysimeter was clipped weekly to a height of 5-8 cm above soil surface and just prior to test substance application. Following application, the grass in each lysimeter was cut during site visits to ensure grass was not longer than 4 cm above the lysimeter side-walls, or 10 cm above ground surface. When cut, the grass clippings remained within the lysimeter boundary, but were not prevented from being removed by natural elements (i.e., wind, wildlife). The test plot area pesticide use history demonstrated that no other chemicals with related chemistry were applied within 2 years. No maintenance pesticides were applied to the plot area throughout the duration of the study. No fertilizer application was required.

Climate and Irrigation

Rainfall, air temperature, relative humidity, wind speed and direction, solar radiation and soil temperature at depths of 2.5, 10 and 100 cm were monitored at the test site using a Campbell 21XL datalogger. The lysimeters were irrigated by hand held sprayers on a weekly basis throughout the 4-month study to ensure that total precipitation plus irrigation amounts were at least 130% of the local 30-year normal rainfall for the same period. Surplus rainfall plus irrigation exceeding the weekly target was not carried over into subsequent weeks. During the weeks when leachate samples were collected, the weekly irrigation of each lysimeter was performed after the leachate samples were collected.

Test and Tracer Substances

The test substance for this study was Lontrel Turf & Ornamental Herbicide, an end-use formulation containing clopyralid as the monethanolamine salt. Prior to application, the test substance assay revealed a clopyralid amine-salt concentration of 42.0% (by weight), and the clopyralid acid equivalent (ae) concentration of 31.8% ae (or 369.1 g ae L^{-1}). On the day of application (8-June-2000), the test substance was mixed with water, after which a sample was collected revealing a clopyralid (ae) concentration of 0.20 mg (ae) mL^{-1}.

A conservative, non-reactive tracer substance, potassium bromide (KBr), was applied in order to track the movement of solute through the soil profile, and for comparison with the center of mass movement of the test substance. The center of mass of the test material and tracer was determined similar to methods outlined in (23). The tracer substance was mixed with carrier (water) on the day of application. Prior to application, a sample was collected for assay of Br^- ion in treatment solution.

Test Substance Application

The eight treatment lysimeters were treated with the test substance at a target rate of 280 g ae ha^{-1} (0.25 lb. ae $acre^{-1}$). A layer of 4-mil plastic sheeting was placed around each of the eight test lysimeters to ensure that no cross contamination occurred during the application activities.

On the day of application, 8-June-2000, at 11:36 a.m., the test solution was prepared and applied to each of the eight treatment lysimeters at a target rate of 280 g (ae) ha^{-1}. A specially designed application apparatus was placed on top of each test lysimeter. The application apparatus consisted of a flat metal plate containing 20 holes which were configured to ensure a uniform application. In treating each lysimeter, a calibrated pipette was used to inject 0.5 mL solution of clopyralid into each of 20 holes, yielding a total of 10 mL of test solution applied per lysimeter. The test solution was assayed post-application and contained an average of 199 µg (ae) mL^{-1} (ppm). Therefore, assuming a lysimeter diameter of

30.5 cm and an area of 729.66 cm^2, each lysimeter was treated at a theoretical application rate of 272.73 g (ae) ha^{-1}, or 97.4% of the target 280 g (ae) ha^{-1}.

After completing the application of the test substance to all of the treatment lysimeters, the lysimeters were allowed to dry for 30 minutes. At the end of this drying period, the tracer substance was applied at a rate of 112 kg Br⁻ ha^{-1} (100 pounds Br⁻ acre^{-1}).

The application rate was verified by sampling surface soil from one of the eight treated lysimeters designated for soil sampling only on the day of application. Three soil cores were collected with the use of a 5 cm diameter split-spoon probe inserted to approximately 8 cm soil depth. The cores were then combined to produce one composite sample, which was shipped on dry ice by overnight express delivery to the analytical facility.

Lysimeter Sampling

Of the nine lysimeters, four were designated for test and tracer substance treatment, and subsequent sampling for leachate only. Four were designated for treatment and sampling for leachate, but only until a given soil core sampling event, at which time, the lysimeter was decommissioned from any further sampling. One was designated for no treatment to provide control samples. At the end of the study, three soil cores were collected from each of the four leachate-only units to aid in calculating Br⁻ tracer material balance. The sampling period was four months, 8-June-2000 through 11-October-2000.

Leachate and soil samples were placed on dry-ice within 10 minutes. of collection, and shipped on dry ice by air express delivery to the analytical facility.

Leachate Sampling

Prior to tracer or test substance application, leachate was sampled from all nine lysimeters, and designated as pre-application samples. Following tracer and test substance application, leachate was collected from both leachate-only and leachate/soil core units on a bi-weekly basis. Leachate samples were collected by connecting a self-priming, variable speed peristaltic pump to the end of dedicated, non-reactive tubing placed in the lysimeter reservoirs. The samples were withdrawn from the reservoir and placed directly into sample containers. A duplicate sample was split from each sample to produce one sample for clopyralid analysis, and one for Br⁻ tracer analysis. At each sampling event, the total volume of leachate drainage was measured for each lysimeter.

Soil Sampling

The four leachate/soil core units were used to sample soil at various intervals for the purpose of determining soil half-life of clopyralid, and for tracking Br⁻ tracer movement. Of the four leachate/soil core units, one was sampled for soil on the day of application, 0 days after application (0 DAT), after test and tracer substances were applied to all eight treatment lysimeters. The remaining three leachate/soil core units were sampled for soil at 14, 28 and 56 DAT. At each sampling event, three 90 cm cores were collected from one leachate/soil core unit, sectioned into 15 cm segments, and combined to produce one composite sample per 15 cm segment. The surface 15 cm samples included turfgrass. Each hole left after probing was replaced with a 5 cm diameter PVC capped pipe. Sample splitting for clopyralid and Br⁻ tracer analysis occurred at the analytical facility. The four leachate-only units were sampled for soil at the end of the study (125 DAT) in the same manner as described above for the leachate/soil core units.

Analytical Methods

Clopyralid was found to be stable in many different soils stored frozen for up to 378 days (*15, 16, 17, 18*) and in water for 330 days (*19*). For this study, all samples were stored frozen for a maximum of approximately 30 days prior to preparation and analysis at ABC Laboratories, Inc., Columbia, Missouri. Soil samples were ground using a hammer mill equipped with a 5 mm screen prior to sub sampling and analysis. Soil samples were analyzed for clopyralid according to a validated gas chromatography (GC) method using electron capture (EC) with a limit of detection at 0.003 µg g^{-1} (*21*). In brief, residues of clopyralid were extracted from soil using a 90% acetone/10% 1.0N HCl acid solution. Following evaporation of the acetone, the sample is diluted with 1.0N HCl and purified using C_{18} and alumina solid-phase extraction (SPE). The eluant from the alumina SPE is evaporated to dryness and then the residue reconstituted and derivatized with acidified 1-propanol. Following derivatization, the 1-propanol is evaporated from the derivatizing solution, and the clopyralid 1-propyl ester derivative partitioned from an aqueous NaCl solution into hexane containing clopyralid 1-butyl ester as an internal standard. A portion of the hexane extract is then analyzed by capillary GC with electron capture detection (ECD). Leachate samples were analyzed for clopyralid (ae) according to a validated gas chromatography method using electron capture with a limit of detection at 0.366 µg L^{-1} (*20*). Leachate and soil samples were analyzed for Br⁻ using ion chromatography analytical methodology with limits of detection at 0.104 mg L^{-1} and 1.10 µg kg^{-1}, respectively (*22*). Appropriate quality control measures were

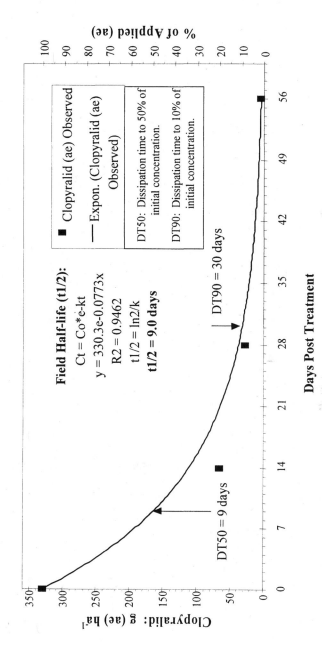

Figure 1. First-order field half-life (DT_{50}) and DT_{90} of clopyralid in soil.

observed, utilizing controls, fortified standards, recoveries, and statistical methods.

Results and Discussion

Climatological Conditions

Rainfall plus irrigation amounted to 117% of normal, for the first month post treatment, 178% of normal for the second month and 196% of normal precipitation for the third month post treatment. For the four-month test period, rainfall and irrigation amounted to 552 mm (21.7 in), or 173% of normal precipitation.

Dissipation and Mobility of Clopyralid in Soil

The initial (0 DAT) concentration of clopyralid in soil was 0.398 μg g^{-1}, equivalent to 330 g (ae) ha^{-1} or 121% of theoretically applied 273 g (ae) ha^{-1}. Clopyralid dissipated rapidly to 10% of the concentration initially applied by 30 DAT, with a first-order field half-life of 9 days (r^2 = 0.95), based on nonlinear regression of untransformed data (Figure 1). This field half-life is within the 7-66-day range of field half-life observed in field dissipation studies (Table I). In addition to microbial degradation, potential offsite movement of grass clippings remaining after weekly clipping could also have contributed to the observed dissipation rate.

Maximum depth of downward movement of clopyralid was limited to the 0-15 cm soil profile depth (Figure 2). The low mobility of clopyralid can be explained in part by its increased sorption and with increased soil contact time (1). An independent estimate of sorption was obtained by calculating an apparent field distribution coefficient (K_d) through time as detailed by van Wesenbeeck, et al (23). Apparent field K_d was determined by comparison of clopyralid and Br$^-$ tracer center of mass (COM) at 14-, 28- and 56-DAT (Figure 2) and compared with the 30-day batch equilibrium (laboratory-derived) K_d in a Commerce loam (24) (Figure 3). Early timepoint estimates of apparent field K_d were less reliable due to the effect of low spatial resolution, and thus difficulty in separating the true COM of Br$^-$ and analyte of interest in the soil profile. The increase in apparent field K_d to 0.17 mL g^{-1} after 28 days compares very well to the 30 day laboratory-derived K_d of 0.19 mL g^{-1} (24). Apparent field K_d observed here and with another low K_{oc} herbicide (23) confirm laboratory observations about decreased mobility with time.

These results further point out that 1-day batch equilibrium (laboratory-derived) soil sorption (K_d) results reflect extreme worst-case leaching conditions that only exist in the environment immediately following application. For this reason, the apparent field K_d or 30-day aged laboratory results (~0.19 mL g^{-1}) could be utilized in environmental fate computer simulations to more accurately predict the environmental fate of clopyralid for other than extreme-worst-case conditions (immediately after application). This approach was confirmed by a computer simulation using the USEPA approved PRZM model which revealed that when K_{oc} values of 50-100 mL g^{-1} (K_d ~0.20-0.60 mL g^{-1}) were used as input, the simulated results were in best agreement with the behavior of clopyralid observed in the field (*25*). These observations point out that simulation models such as LeachP and PRZM could be improved by incorporating changing soil sorption with time.

Bromide Tracer in Leachate

Bromide tracer in leachate began to be observed by 56-DAT (Figure 4), and increased to a maximum of 17.5 ppm, or 5% of applied mass, by 112 DAT after 264 mm of cumulative leachate drained through (Figure 5). Cumulative leachate at the end of the 125-day study was 310 mm, or 56% of rainfall + irrigation, indicating that ample opportunity existed for mass solute movement through the soil profile, especially during the first 14 days post treatment when leachate volume was 100% of that received as rainfall and irrigation (Figure 5).

Clopyralid in Leachate

Although Br$^-$ tracer breakthrough and leachate water balance (56% of precipitation) indicated ample opportunity for leaching, clopyralid remained undetected (detection limit 0.4 ug L^{-1}) in leachate samples throughout the entire 125-day test period (Figure 6).

Conclusions

The dissipation of clopyralid after post emergence application to turfgrass at the labeled use rate was rapid, with a first-order field half-life of 9-days. The dissipation half-life observed in this study was within the range of half-life (7-66 days) observed in field dissipation studies. The rapid dissipation rate and sorption characteristics observed in this study are consistent with existing laboratory data generated for clopyralid. Although ample opportunity existed for

226

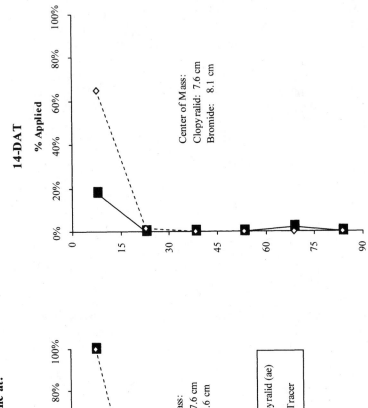

Clopyralid & Tracer Soil Profile at:

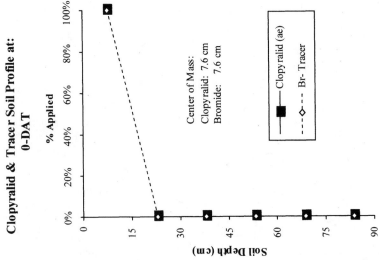

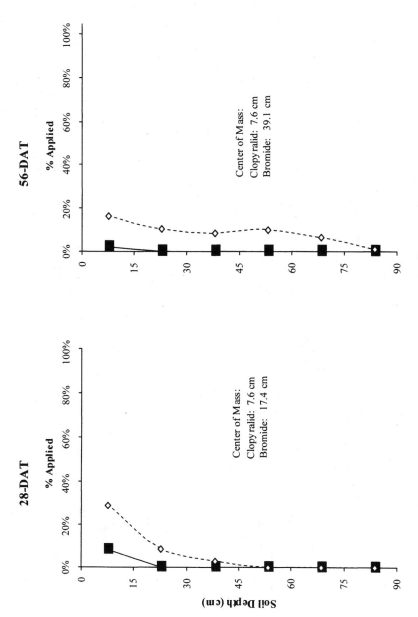

Figure 2. Clopyralid and bromide tracer movement through the soil profile at 0, 14, 28, and 56 Days After Treatment (DAT).

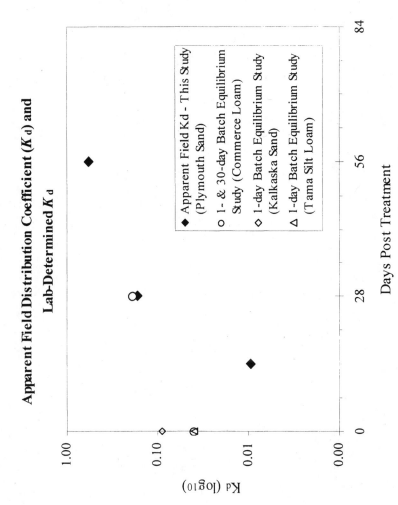

Figure 3. Apparent field K_d of clopyralid vs. lab measured K_d over time.

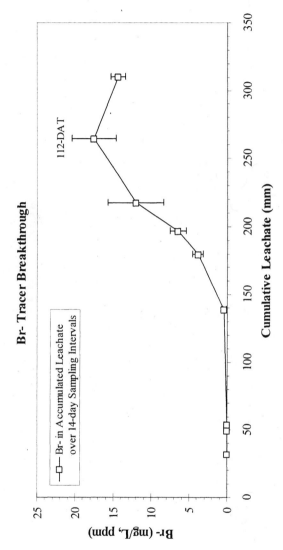

Figure 4. Bromide concentration in leachate as a function of cumulative

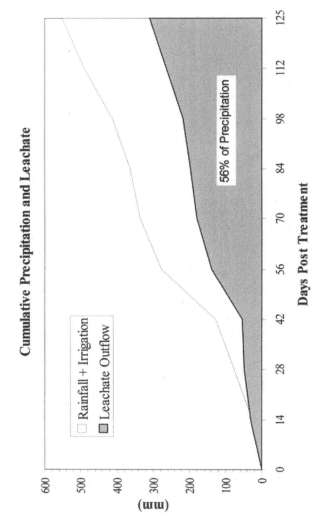

Figure 5. Cumulative precipitation and lysimeter leachate as a function of time.

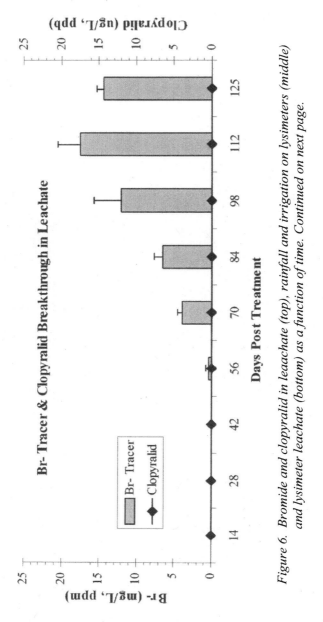

Figure 6. Bromide and clopyralid in leaachate (top), rainfall and irrigation on lysimeters (middle) and lysimeter leachate (bottom) as a function of time. Continued on next page.

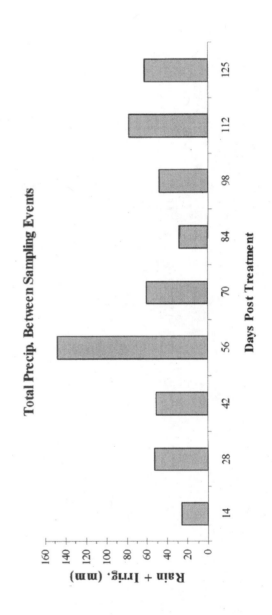

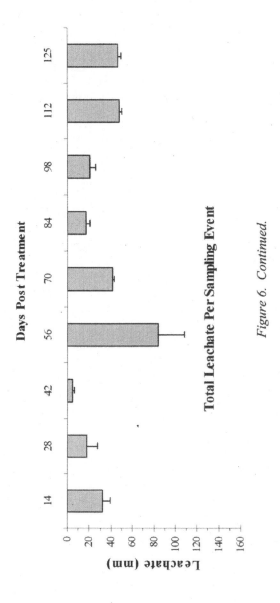

Figure 6. Continued.

234

leaching, and bromide tracer breakthrough was well established after 112 days post treatment, clopyralid remained undetected in leachate throughout the entire 125-day study. Therefore, under typical irrigated turfgrass practices, on a highly vulnerable sand soil, clopyralid is not expected to leach into ground water.

Although some laboratory derived physico-chemical properties of clopyralid (e.g., water solubility and soil sorption) are characteristic of pesticides having a potential to leach, this study demonstrated that leaching potential in the field, under turfgrass use conditions, is extremely low, being offset by 1) rapid and complete microbial degradation, 2) plant interception/uptake, and/or 3) increased soil sorption (K_d) with residence time in soil. It is important for regulatory decision makers to consider these factors when evaluating the potential impacts to surface and ground water resulting from the use of clopyralid and similar compounds for safe use in turfgrass environments.

References

1. *The Pesticide Manual*; Editor, C.D.S. Tomlin; 12th Edition; British Crop Protection Council: Alton, Hampshire, UK, 2000; pp 193-194.
2. Behl, E.; Eiden, C. In *Groundwater Residue Sampling Design*; American Chemical Society: Washington, DC, 1991; pp 27-46.
3. Winton, K.; Weber, J. *Weed Tech.* **1996**, *10*, 202-209.
4. Bergstrom, L.; McGibbon, A.S.; Day, S.R.; Snel, M. *Environ. Toxicol. Chem.* **1991**, *10*, 563-571.
5. Brumhard B.; Bergstrom, L.F.; Snel, M.; Fuhr, F.; Baloch, R.I.; Trozelli, H. *Med. Fac. Landbouww. Rijksuniv. Gent.* **1991**, *56/3a*, 887-900.
6. McGibbon, A.S.; Day, R.S.; Portwood, D.; Bergstrom, L. *GHE-P-2109*, 1990. Unpublished report of Dow AgroSciences.
7. Brumhard B.; Fuhr, F.; Snel, M.; Baloch, R.I. In *Lysimeter Studies of Pesticides in the Soil;* BCPC Monograph No. 53: Alton, Hampshire, UK, 1992; pp. 103-114.
8. Baloch, R. I.; Brumhard B.; Fuhr, F. *GHE-P-2580, 1992.* Unpublished report of Dow AgroSciences.
9. Baloch, R.I.; Brumhard, B; Fuhr, F. *GHE-P-2908*, 1992. Unpublished report of Dow AgroSciences.
10. Jackson, R; Dust, M. *GHE-P-4037*, 1994. Unpublished report of Dow AgroSciences.
11. Reeves, G.; Schnoeder, F. *GHE-P-10833, 2004.* Unpublished report of Dow AgroSciences.
12. Roberts, D.W. *Internal memorandum* to Brinkmeyer, R.S. 2001. Dow AgroSciences.
13. Roberts, D.W.; Walker, T.W.; Bussard, J. *GH-C 5164,* 2000. Unpublished report of Dow AgroSciences.

14. Persson, L.; Bergstrom, L. *Soil Sci. Soc. Am. J.* **1991**, *55*, 285-287.
15. Knuteson, J.A., Ervick, D.K.. *GH-C 2671*, 1991. Unpublished report of Dow AgroSciences.
16. Kloppenburg, D.J., Clayson, J.E. *94DWE01*, 1994. Unpublished report of Dow AgroSciences.
17. Roberts, D.W., Phillips, A.M., Blakeslee, B.A., Swanson, M.E.. *GH-C 3812*, 1995. Unpublished report of Dow AgroSciences.
18. Roberts, D.W., Phillips, A.M., Blakeslee, B.A., Siders, L.E. *GH-C 4255*, 1996. Unpublished report of Dow AgroSciences.
19. Phillips, A.M., Arnold, B.H., Robb, C.K.. *GH-C 4961*, 1999. Unpublished report of Dow AgroSciences.
20. Clements, B., Wicks, D. *ERC 95.23*, 1996. Unpublished report of Dow AgroSciences.
21. Harnick, B.J., Olberding, E.L. *GRM 95.01*, 1995. Unpublished report of Dow AgroSciences.
22. Pfaff, J.D. *Method 300.0 – Determination of Inorganic Anions by Ion Chromatography, Revision 2.1.* 1993. USEPA Office of Research and Development, Environmental Monitoring Systems Laboratory, Cincinnati, OH 45268.
23. van Wesenbeeck, I.J.; Zabik, J.M.; Wolt, J.D.; Bormett, G.A.; Roberts, D.W. *J. Agric. Food Chem.* **1997**, *45*, 3299-3307.
24. Woodburn, K. B., French, B.W. *GH-C 1873*, 1987. Unpublished report of Dow AgroSciences.
25. Oliver, G.R. *GH-C 1833*, 1986. Unpublished report of Dow AgroSciences.

Chapter 14

Glyphosate Runoff When Applied to Zoysiagrass under Golf Course Fairway Conditions

Steven K. Starrett[1] and Jamie Klein[2]

[1]Department of Civil Engineering and [2]Graduate Student, Department of Civil Engineering, Kansas State University, Manhattan, KS 66506

The National Golf Foundation estimates an all time high of more than 27.4 million golfers playing at more than 25,000 courses nationwide in 2003, for a combined total of approximately 500 million rounds of golf annually. The operation of these golf courses requires significant input of various pesticides, which can potentially be harmful to the surrounding environment. In 2001, 1.2 billion pounds of pesticides were applied to all land uses in the United States, 44% of which were herbicides. Glyphosate is the primary ingredient of Roundup®, which is the most widely used herbicide by volume in the United States. Glyphosate runoff is therefore an important subject for both existing and newly constructed courses. The objectives of the current research were: (1) to measure glyphosate runoff from zoysiagrass fairways on a golf course following the application of Roundup herbicide, (2) to determine glyphosate runoff concentrations and their resulting effect on the environment, and (3) to provide up-to-date data of research findings on pesticide transport when applied to turfgrass. Previous research on glyphosate runoff from turfgrass has been done on test plots and not full-scale watersheds. In addition, Roundup applications for this study were made by the Colbert Hills Golf Course staff and not prescribed specifically for research purposes. Water quality and quantity monitoring systems were set up on the 115-acre study watershed, which contains a 3-acre detention and irrigation pond. Over 600 water samples were taken from fairway drains, the inlet and outlet of the pond, and the pond itself throughout a three-year study period. Ten of the twenty-three tested samples contained detectable concentrations (>0.10 μg/L) of glyphosate. The maximum observed glyphosate concentration was 5.18 μg/L, which is well below the United States Environmental Protection Agency (USEPA) Drinking Water Maximum Contaminant Level (MCL) of 700 μg/L. These results suggest that Roundup applications made to turfgrass fairways do not cause hazardous levels of glyphosate in downstream surface water.

This study focused on observing the fate of glyphosate following Roundup applications in a turfgrass ecosystem. Kansas State University, in cooperation with Jim Colbert, the Professional Golf Association (PGA), the Golf Course Superintendents Association of America (GCSAA) and various alumni, built a championship golf course (Colbert Hills Golf Course) in Manhattan, Kansas. The golf course opened in 2000, and provided the study area for the research presented here. Water and soil samples were taken from the study watershed for 3 years (2001-2003) during the early operation period. This research is an extension of nutrient runoff and modeling studies completed by Dr. Su (*1*) and Mr. Heier (*2*).

The objectives of this study were:

1. To measure glyphosate runoff from zoysiagrass (*Zoysia japonica* Steud., Meyer cultivar) fairways on a golf course following the application of Roundup herbicide.
2. To determine glyphosate runoff concentrations and the resulting effect on the environment.
3. To provide up-to-date data of research findings on pesticide fate and transport when applied to turfgrass.

This research is unique because the pesticide applications were made as needed and not prescribed for research purposes, and because of the full scale watershed area that was used.

Glyphosate is a non-selective, systemic, post-emergence herbicide registered for use on food and non-food field crops, non-crop areas where total vegetation control is desired and can also act as a plant growth regulator (*3*). Glyphosate [N-(phosphonomethyl) glycine] is generally sold as the isopropylamine salt and applied as a liquid foliar spray. Glyphosate is among the most widely used pesticides by volume, with the majority being used in agricultural settings. In recent years, an estimated 13 to 20 million acres were treated with 18.7 million pounds annually (4). Some common trade names of products that contain glyphosate as the active ingredient are Roundup, Rodeo®, and Accord®.

Glyphosate Health Impacts

With the introduction of glyphosate to the environment, a number of environmental considerations arise, probably the most important being the health implications to those who may come in contact with the chemical. The USEPA RED (1993) fact sheet reported that, in California, glyphosate ranked high among pesticides causing illness or injury to workers, who report numerous incidents of eye and skin irritation from splashes during mixing and loading. Glyphosate is categorized as low in toxicity regarding eye irritation. Scientists who evaluated eye effects from 1,513 reported human ocular exposures to glyphosate products found that the majority experienced no injury (21%) or temporary minor effects (70%). Only 2% of the cases involved symptoms that required medical treatment, while none of the exposures resulted in permanent eye damage or loss of visual acuity (5).

Most of the following toxicity and health information regarding glyphosate will have corresponding LD_{50} or LC_{50} values. LC_{50} is defined as the *lethal concentration*, or the concentration of a potentially toxic substance in an environmental medium that causes death of 50% of the organism population following a certain period of exposure. Likewise, LD_{50} refers to the dose that causes death of 50% of the organism population. These levels should not be reached in order to maintain the current ecosystem in the watershed. Glyphosate is low in toxicity when applied to the skin. The acute dermal LD_{50} for glyphosate in rabbits is >2000 mg/kg (6). A study of 204 human volunteers dermally exposed to an undiluted glyphosate product showed no skin sensitization with any of the participants, confirming that glyphosate is absorbed poorly through the skin (3).

Because glyphosate is not readily volatilized, the risk of inhalation is relatively low. The acute inhalation LC_{50} of glyphosate in rats is >0.81 mg/L. Studies have demonstrated that a very low percentage (5-10% at most) of the total pesticide residue present immediately following application is in dislodgeable form, and that concentrations reduce rapidly with time. Studies of air quality during and directly following application have shown that very low levels of pesticides were present, indicating that inhalation is, at most, a secondary route of potential exposure (7). Ingestion is another potential avenue for human exposure to glyphosate. The acute oral LD_{50} in rats was 4,320 mg/kg. To date, no long term chronic effects have been observed from accidental ingestion of glyphosate. However, there are accounts of five patients dying from intentional ingestion of glyphosate products (8). Glyphosate is poorly absorbed from the digestive tract and is largely excreted unchanged by mammals. Ten days after treatment, there were only minute amounts in the tissues of rats that were fed glyphosate for three weeks (9).

As with acute toxicity, chronic toxicity levels have been determined through animal studies, which are used to make inferences relative to human health. The USEPA has evaluated the reported toxicity data in support of the registration of glyphosate (10). A reproduction study of 3 generations of rats has shown no adverse effects on fertility or reproduction at doses up to 30 mg/kg per day, therefore it is unlikely that glyphosate would produce any reproductive effects in humans (9, 10). Teratogenic or developmental effect studies have also shown no evidence of birth defects in pregnant rats and rabbits that were subjected to dose levels of 3500 mg/kg and 350 mg/kg, respectively (10). Glyphosate tested negative on eight different kinds of bacterial strains and on yeast cells for mutagenic effects (the ability to cause genetic damage) (11). Organ toxicity studies show that glyphosate caused no changes in the rate of body weight gain, in blood, nor in the kidneys or liver at doses up to 500 mg/kg (12). The USEPA has stated that there is sufficient evidence to conclude that glyphosate is not carcinogenic in humans (13).

Drinking water is another possible source of glyphosate exposure to humans. Effects of acute exposure to glyphosate in drinking water may include congestion of the lungs and increased breathing rate. Long term exposure to glyphosate in drinking water at levels above the maximum contaminant level (MCL) may result in kidney damage or reproductive effects. Glyphosate tends to become slightly more toxic when formulated as a commercial product due to the presence of surfactants. The toxicity of the technical product glyphosate and the formulated product Roundup is nearly the same. The USEPA (4) National Primary Drinking Water Regulations are shown in Table I. The MCL is defined as the maximum permissible level of a contaminant in water that is delivered to any user of a public water system. MCLs are enforceable standards established by the USEPA. The health advisory level (HAL) is defined as a non-regulatory health-based reference level of chemical traces in drinking water at which there are no adverse health risks when ingested over various periods of time. Such levels are established for one day, 10 days, and long-term / life-time exposure periods. They contain a wide margin of safety.

Safety precautions and medical treatment procedures are fairly consistent for all glyphosate products. Roundup, the most commonly used glyphosate product, carries the following signal word and statement on its label: "**WARNING –** CAUSES EYE IRRITATION. HARMFUL IF SWALLOWED. MAY CAUSE SKIN IRRITATION". Medical treatment instructions are to treat the symptoms encountered for glyphosate exposure. For eye exposure, flush with plenty of water for at least 15 minutes and get medical attention. For skin exposure, flush

Table I. National Primary Drinking Water Regulations from the United States Environmental Protection Agency (USEPA) for Glyphosate

Exposure Guideline	Limit (mg/liter)	
MCL (maximum contaminant level)	Lifetime:	0.7
	1-10 day:	20.0
HAL (health advisory level)	Longer-term:	1.0
	(children)	

skin with plenty of water. In case of emergency, call the local poison control center for advice (*10*).

Glyphosate toxicity to wildlife is also an important issue to consider. Glyphosate has been found to be practically nontoxic to birds, honeybees, fish and aquatic invertebrates. Toxicity information is shown in Table II (*3*). Based on its water solubility, glyphosate is not expected to bioconcentrate in aquatic organisms. The BCF, or bioconcentration factor, for glyphosate in fish exposed for 10-14 days was 0.2 to 0.3 (*4*). The BCF is defined as the ratio of chemical concentration in the organism to that in surrounding water. Glyphosate can also cause harm to wildlife by altering their natural habitat. Glyphosate concentrations typically found in runoff pose no risk to the surrounding wildlife.

Table II. Ecological Toxicity Information for Glyphosate from the USEPA's Reregistration Document

Birds	Honeybees	Fish	Aquatic Invertebrates
$LD_{50} > 2000$ mg/kg	$LD_{50} > 100$ μg/bee	$LC_{50} > 24 - 140$ mg/L	$LC_{50} > 780$ mg/L

Environmental Fate of Glyphosate

The fate of glyphosate is influenced by its numerous interactions with the surrounding environment. The most notable interactions take place with water, soil, air and the treated vegetation. The fate of glyphosate through these different modes is highly dependent on its physical properties.

Glyphosate is a non-selective, systemic, post-emergence herbicide that is most often applied by spraying for total vegetation control. Drifting spray that comes in contact with non-target plants may injure or kill them. Although no known plants are naturally resistant to glyphosate, scientists have genetically altered some crop plants allowing glyphosate to be used as a selective herbicide

(*14*). Common use rates of glyphosate range from 0.35 to 0.44 kg of active ingredient per ha (*10*). Glyphosate is absorbed across the plant surface and is translocated throughout the plant (*15*). Virtually all of the glyphosate delivered to the plant originates from the plant surface. Ghassemi et al (*16*) found that less than one percent of the glyphosate found in the soil is absorbed by the roots.

Once absorbed into the plant, glyphosate tends to accumulate in meristematic plant regions, or where cells are actively dividing (*14, 17*). The target of glyphosate is the inhibition of the enzyme 5-enolpyruvylshikimate-3-phosphate synthase (EPSPS), which is a chloroplast-localized enzyme in the shikimic acid pathway found in plants and microorganisms (*14, 18*). Inhibition of EPSPS prevents the production of chorismate, which is required for the biosynthesis of essential aromatic amino acids (*15*). Unlike many contact herbicides, the effects of glyphosate may not be visible for up to a week. Plants exposed to glyphosate will exhibit stunted growth, loss of green coloration, leaf wrinkling or malformation and tissue death (*15, 19*).

Glyphosate is most often applied as a spray, and is removed from the atmosphere by gravitational settling (*4*). Volatilization is the process by which the liquid or solid that reaches the plant surface transforms into a gas. The tendency for volatilization to occur is directly related to vapor pressure, temperature and air movement (*20*). The vapor pressure for glyphosate is very low; therefore the fate pathway through volatilization is virtually nonexistent. In addition, glyphosate's low Henry's Law Constant indicates a tendency to partition into water versus air (*21*).

Glyphosate has the potential to reach surface water through accidental spraying, spray drift, or surface runoff. Glyphosate runoff usually occurs because it adsorbs to soil particles which become suspended in runoff water (*3*). Glyphosate is highly soluble in water (11,600 ppm at 25°C), with an octanol-water coefficient (log K_{ow}) of 3.3 (*22*). Studies conducted by the USEPA have shown that glyphosate is stable to photodegradation under natural sunlight, and to hydrolysis (*3*). Therefore, once in surface water, degradation occurs primarily with sediment deposition and microbial decomposition, with minute amounts being broken down by water and sunlight (*3, 4*). Ghassemi et al (*16*) concluded that the rate of degradation in water is slower than in soil because fewer microorganisms are available. USEPA half-life tests using water from natural sources showed a range for glyphosate of 35 to 63 days. (*21*).

Glyphosate that makes its way to the soil will readily adsorb and be biodegraded into aminomethylphosphonic acid (AMPA) and carbon dioxide. Glyphosate's high soil adsorption coefficient (K_d=61 g/cm^3) and very low octanol-water coefficient (log K_{ow}=3.3) suggest low mobility and very slight leaching potential (*21*). Once adsorbed, glyphosate's primary route of decomposition in the environment is through microbial degradation in soil (*15*). The herbicide is inactivated and biodegraded by soil microbes at rates of degradation directly related to microbial activity in the soil (*23*). The biological degradation process is carried out under both aerobic and anaerobic conditions by soil microflora at rates dependent on population types (*10*). The average aerobic and anaerobic half lives of glyphosate in soil are 7 days and 8 days, respectively (*3*).

AMPA is the primary metabolite of glyphosate. The degradation of AMPA is generally slower than that of glyphosate, possibly because it adsorbs more strongly onto soil and/or it may be less likely to permeate the cell walls or membranes of soil microorganisms (*10*). The Monsanto Company has stated that Rueppel et al (*24*) proposed the possible formation of formaldehyde from AMPA in 1977. Since that possibility was suggested, extensive studies have failed to detect formaldehyde as a decomposition product of either glyphosate or AMPA (*25*). In 1997 the USEPA stated that AMPA is not of toxicological concern, regardless of its levels in food (*26*). Research conducted by Stratton and Stewart (*27*) has shown that glyphosate generally had no significant effect on the numbers of bacteria, fungi, or actinomycetes microorganisms in the soil. In addition, the strong soil adsorption characteristics of glyphosate limit its phytotoxicity to other plants through the soil (*19*). Another benefit of the strong adsorption is limited leaching potential; glyphosate is not expected to move vertically below the six inch soil layer (*3*). Jonge (*27*) determined that the risk for glyphosate leaching from the topsoils seems to be limited to conditions where pronounced macropore flow occurs shortly after application.

Materials and Methods

Runoff samples were collected by Isco® 3700 (Teledyne Isco, Inc.) automated water samplers at the inlet and outlet of the subwatershed detention pond (Figure 1). The samplers were activated by a liquid actuator that detected a 10-cm rise in the stream level, therefore only sampling during runoff events. Because the stream was intermittent, significant precipitation was often required to actuate the sampler. Approximately ten such runoff events were expected per year. The samplers were programmed to take 4 samples per bottle at 15-minute intervals, therefore filling 1 bottle per hour. Twenty-four bottles inside each Isco sampler permitted 24 hours of composite sampling. Between samplings, the bottles were rinsed with acetone and distilled water to prevent cross-contamination. A total of 211 runoff samples were taken from the inlet, and 125 samples from the outlet, throughout the study period (Table IV).

Water samples were also taken from three separate fairway drains on hole number 9 (Figure 1). Table III shows the approximate drainage areas and the percentages of each cover type drained. The fairway drains are connected and eventually drain into the detention pond (Figure 1). To take samples from the drain, a glass bottle was developed by Starrett (*28*) that automatically seals when the bottle is full. This ensured that the samples taken would not be contaminated by either dilution or excessive loading from continued runoff. There were a total of 61 samples taken from three separate fairway drains throughout the study period (Table IV).

Table III. Drainage Areas and Percent of Fairway and Rough Drained for Each of the Three Fairway Drains on Hole #9 at Colbert Hills Golf Course

| | Drainage Area (acres) | Percent of Cover Type Drained | |
		Zoysia fairway (Zoysia)	Fescue Rough (Festuca)
Fairway Drain #1	0.37	70	30
Fairway Drain #2	0.30	70	30
Fairway Drain #3	0.46	63	37

Water samples were also taken directly from the pond once every month. 'Pond 1' and 'Pond 2' were used to designate two separate areas within the same body of water. Pond 1 is a more shallow area with more vegetative cover near the inlet of the pond, and Pond 2 is the deeper area near the outlet. There were seven specific locations in the pond where samples were taken at three different depths (25%, 50%, and 75% of total depth). Glass containers were placed on a measuring rod for sample collection so that samples would be collected at a consistent depth. A total of 208 samples were taken throughout the study period from different locations and depths of the pond (Table IV). All of the collected water samples were stored in a freezer until testing.

Results

Due to budgetary limitations, twenty-three water samples were tested by the United States Geological Survey (USGS) lab in Lawrence, KS for glyphosate and AMPA (Table V). Precipitation recorded for each sampling occasion is provided in Table VI. Although glyphosate concentrations were the primary concern, the USGS testing procedure for glyphosate also screens for AMPA, the primary metabolite of glyphosate. Samples from each location, including fairway drains, inlet to pond, outlet from pond and from the pond itself were tested for glyphosate.

An emphasis was placed on samples from fairway drains taken early in the year because they were closest to the Roundup application date. The limit of detection for the testing process used was 0.10 μg/L. The majority of samples with detectable concentrations of glyphosate and AMPA were taken from

Table IV. The Number of Samples Taken from the Study Area at Colbert Hills Golf Course Throughout the Study Period of 2001-2003

Location	2001	2002	2003	Total
Pond 1	24	9	19	52
Pond 2	64	38	54	156
Inlet	95	79	37	211
Outlet	87	1	37	125
Fairway Drain #1	8	7	6	21
Fairway Drain #2	9	4	8	21
Fairway Drain #3	8	5	6	19
Pond Sediment	5	7	0	12
Annual Total	300	150	167	617

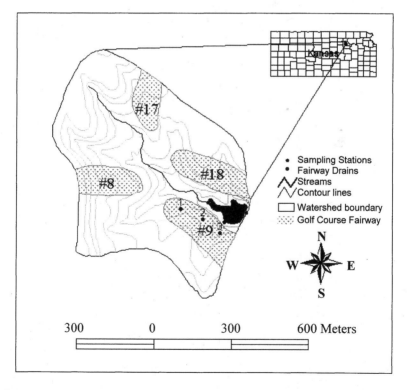

Figure 1. Location of fairway drains on hole #9 at Colbert Hills Golf Course. The fairway drains eventually connect and drain into the detention pond. (See page 4 of color inserts.)

Table V. Glyphosate and AMPA Concentrations Found in Samples Taken from Colbert Hills Golf Course (Manhattan, KS

Location	Collection Date	Glyphosate (µg/L)	AMPA (µg/L)	Time after app. (days)
Drain #1	10/18/01	<0.10	<0.10	----[a]
Drain #1	10/18/01	<0.10	<0.10	----[a]
Drain #1	05/07/02	1.23	5.41	62
Drain #1	05/12/03	0.33	1.03	60
Drain #1	05/21/03	3.25	6.11	69
Drain #2	05/07/02	5.18	21.60	62
Drain #2	06/20/02	0.42	1.21	106
Drain #2	05/12/03	0.59	1.55	60
Drain #3	10/18/01	<0.10	<0.10	----[a]
Drain #3	05/07/02	0.47	1.24	62
Drain #3	05/12/03	1.02	2.49	60
Drain #3	07/02/03	0.63	0.80	111
Drain #3	07/23/03	<0.10	0.64	132
Inlet	04/29/02	0.16	<0.10	54
Inlet	05/05/02	<0.10	<0.10	60
Inlet	05/10/02	<0.10	<0.10	65
Inlet	04/23/03	<0.10	<0.10	41
Inlet	05/24/03	<0.10	<0.10	72
Pond 1 (0.75')	04/16/03	<0.10	<0.10	34
Pond 2 (1.0')	04/16/03	<0.10	<0.10	34
Pond 2 (6.0')	08/06/03	<0.10	<0.10	146
Pond 2 (8.0')	08/06/03	<0.10	<0.10	146
Outlet	05/24/03	<0.10	0.26	72

a* = Roundup was not applied in 2001.
NOTE: Pond samples were taken independently from precipitation events.

Table VI. Precipitation Data for Samples Taken from Colbert Hills Golf
Course (Manhattan, KS)

Collection Date	Precipitation (in.)	Average intensity (in/hr)	Maximum intensity (in/hr)
10/18/01	0.70	0.10	0.20
10/18/01	0.70	0.10	0.20
05/07/02	1.40	0.03	1.10
05/12/03	1.40	0.47	0.30
05/21/03	0.50	0.06	0.10
05/07/02	1.40	0.03	1.10
06/20/02	0.21	0.02	0.10
05/12/03	1.40	0.47	0.30
10/18/01	0.70	0.10	0.20
05/07/02	1.40	0.03	1.10
05/12/03	1.40	0.47	0.30
07/02/03	1.02	0.13	0.44
07/23/03	0.61	0.21	0.53
04/29/02	0.30	0.10	0.10
05/05/02	1.20	0.60	1.10
05/10/02	0.60	0.30	0.30
04/23/03	1.40	0.14	0.40
05/24/03	0.30	0.10	0.10
04/16/03	----	----	----
04/16/03	----	----	----
08/06/03	----	----	----
08/06/03	----	----	----
05/24/03	0.30	0.10	0.10

NOTE: Pond samples were taken independently from precipitation events.

fairway drains. Higher concentrations were generally observed earlier in the year, closer to the time of Roundup application. Total precipitation during the runoff event, average and maximum rainfall intensities, and the time between Roundup application and sample collection were all important factors of glyphosate concentrations found in runoff.

Discussion

In addition to Roundup application and precipitation data, other factors affecting the concentration of glyphosate in runoff water include turf cover type and characteristics, the existence of thatch and buffer strips, and the soil moisture prior to runoff events.

Zoysiagrass is a warm season sod-forming perennial turfgrass typically found in the southern half of the United States. Zoysiagrass grows well in a wide variety of soils from sands to clays. Zoysiagrass conserves moisture more efficiently than other species and is extremely drought tolerant, with deep root systems enabling it to more effectively extract water from greater depths. Close and frequent mowing results in the best turf conditions and prevents the over-accumulation of a thatch layer (29).

Colbert Hills practices frequent and close mowing procedures and therefore thatch accumulation is minimal, although a small thatch layer is present. Fairway drains are surrounded by zoysiagrass, which acts as a filter for runoff water. Additionally, native grass species surround the fairways acting as buffer strips to intercept and filter any overland flow that exits the fairways. Because irrigation guidelines of replacing evapotranspiration (ET) are followed, the soil moisture prior to runoff events should not be significantly greater beneath the zoysiagrass fairways and tee boxes. Therefore, these factors are acting together to reduce (or at least not increase) glyphosate runoff concentrations.

The concentrations of glyphosate found in the samples varied because of multiple factors. From previous studies, the primary factors that influenced glyphosate concentrations in runoff were the Roundup application rate and the amount of time between the application and sampling dates. In addition to these factors, the percent of treated area within the drainage area, and precipitation factors such as total precipitation and maximum intensity were expected to influence glyphosate concentrations.

Variation of Glyphosate Concentrations by Location

The majority of samples with detectable concentrations of glyphosate were taken from one of the three fairway drains. The samples taken from the fairway drains contained significantly greater concentrations than the inlet, outlet, and pond samples. Nine of the thirteen samples taken from the fairway drains during the years of 2002 and 2003 (when Roundup was applied) contained glyphosate at a concentration greater than 0.10 μg/L, while ten of the thirteen samples contained AMPA at a concentration greater than 0.10 μg/L. The sample with the highest concentration observed throughout the study period was taken from fairway drain #2 on May 7, 2002, which was 62 days after Roundup application. The glyphosate concentration of this sample was 5.18 μg/L, which is significantly lower than the 700 μg/L limit for the USEPA established maximum contaminant level (MCL). This sample was taken as part of a 1.40 inch precipitation event with a maximum intensity of 1.10 inches per hour.

There are three likely reasons that the fairway drain samples contained higher concentrations than samples taken elsewhere. First, samples taken from fairway drains were nearest to the source of glyphosate, the zoysiagrass fairways where the Roundup applications were made. This would obviously promote higher concentrations of glyphosate in runoff water as compared to the inlet or outlet of the pond. Second, fairway drain samples were taken as a result of runoff through and over the dense zoysiagrass turf where interaction with soil is limited. Because of this, the glyphosate would not have as much interaction with the soil, and therefore less adsorption and degradation would occur. In addition, the fairway runoff could contain a greater percentage of zoysiagrass clippings which contain Roundup and glyphosate. Finally, the fairway drains each had a drainage area of less than one-half of an acre, with approximately 70% being comprised of zoysiagrass fairways that were treated with Roundup. A comparison with the inlet area to the pond, which drained approximately 115 acres of which only 6% was zoysiagrass, indicated the potential for much greater dilution of glyphosate exiting the inlet area.

Five samples taken from the inlet into the pond were tested by the USGS lab. The one sample that contained glyphosate at a concentration above the detection limit of 0.10 μg/L was taken on April 29, 2002. This was earlier in the year than any of the fairway drain samples, which might explain why a detectable concentration was present in this sample. As previously mentioned, the primary reason for the inlet samples containing undetectable concentrations of glyphosate was dilution. The drainage area for the inlet sampling location was 115 acres with approximately 6% of that being zoysiagrass fairways that were treated with Roundup.

Four samples that were taken directly from the detention and irrigation pond were tested for glyphosate. Two of the samples were taken early in the year (in April, 2003), and the other two were taken later in August, 2003. None of the four samples contained glyphosate at a concentration greater than 0.10 μg/L. There were two primary reasons why low concentrations might be expected within the pond. First, significant dilution occurred when runoff from the watershed mixed with the water present in the 3 acre pond. The pond is used as storage for irrigation water. So, further dilution also occurred due to the addition of purchased water that was pumped into the pond for irrigation purposes. Second, the reduced runoff velocity paired with readily available suspended solids promoted glyphosate adsorption to soil particles and settling.

One sample taken from the outlet of the detention pond was tested for glyphosate. While the concentration of glyphosate was below the detectable limit of 0.10 µg/L, an AMPA concentration of 0.26 µg/L was present. Low concentrations of glyphosate were expected from the outlet for the same reasons mentioned above for the pond samples. Because water from the pond is used for irrigation, it was not favorable for water to be lost through the outlet. Therefore, outlet discharge events would likely occur only after significant or frequent precipitation events. This was the case with the tested sample which was taken 12 days after a 1.4 inch precipitation event. In these cases, it was likely that due to the pond being full, the amount of water entering approximately equaled the amount exiting, which left less time for dilution, adsorption, and settling of glyphosate.

Conclusions

Glyphosate concentrations found in tested runoff samples from the 115 acre study watershed, following annual applications of Roundup, were much lower than associated health standards. The factors present during the study period effectively maintained glyphosate concentrations at acceptable levels at fairway drains, and at undetectable levels within the detention pond. This information should add to the knowledge base of pesticide runoff from golf courses, and be used to help make informed, scientifically based decisions.

Influencing Factors and Fate of Glyphosate

A variety of factors were responsible for the amount of glyphosate that entered into the surface water following Roundup applications. The first factor was the application rate of 20-28 oz/acre of Roundup, which contains 41% glyphosate. These rates were similar to those that Edwards et al (29) observed, and are typical rates for winter-dormant zosiagrass turfgrass use.

The physical characteristics of the golf course, including sample location, turf type, cutting procedures and height, and the existence of a thatch layer and buffer strips were also important. The most significant change due to sample location was the percentage of drained area that was treated with Roundup. Samples taken from fairway drains were far less diluted than samples taken at other locations. The zoysiagrass, thatch layer and buffer strips slowed the runoff velocity and acted as a filter to reduce erosion and therefore glyphosate transport into the downstream surface water. These factors helped to reduce glyphosate runoff and are representative of most golf courses with zoysiagrass fairways.

The final factor that influenced glyphosate runoff was the general hydrology of the study area, including precipitation, ET, irrigation and the presence of a detention pond. Precipitation throughout the study period was fairly consistent, with an overall average annual precipitation of 34 inches. In addition, the greatest intensity observed during the study period was slightly greater than the one-year return period for a one-hour duration storm event. The irrigation practices used at Colbert Hills also helped to reduce glyphosate concentrations in runoff. By using ET as the guideline for irrigation, the amount of water applied can closely match the amount needed by the turf. Therefore the soil moisture is not greatly increased prior to precipitation events. This allows more infiltration to take place during rainfall and thus decreases the runoff volume and pesticide transport. Finally, the detention and irrigation pond reduced glyphosate concentrations by allowing settling and dilution. Glyphosate becomes mobile through adsorption to soil particles which will settle out of the water upon entering the pond. The glyphosate that doesn't settle out of the water will be significantly diluted by the time it exits the pond.

Statistically significant relationships could not be formed because of the limited number of tested samples and relatively low variation of the independent variables used. The concentration of glyphosate in runoff at locations other than the fairway drains was effectively decreased by a combination of degradation and dilution, and is therefore not expected to be present in hazardous concentrations downstream.

Health Standards and Decision Making

Tested sample results showed that the levels of glyphosate found in runoff from the golf course were far lower than any health standards. The maximum concentration of glyphosate found in any sample was 5.18 μg/liter. The lowest limit of the USEPA National Drinking Water Regulations was the Maximum Contaminant Level (MCL) of 700 μg/liter. This is the maximum concentration at which glyphosate can be present in a public water system. Therefore, there was no health risk associated with the glyphosate levels found in runoff throughout the study period.

This study should therefore show that not all pesticide applications have negative impacts on the surrounding environment. Referring back to the Canadian pesticide bans, decisions and legislation regarding pesticide usage should be based on scientific evidence and not on emotions. For example, after reviewing the mechanisms by which glyphosate is degraded naturally in the environment, and after observing that concentrations found in runoff are much lower than any health standards, it is clear that proper Roundup applications on

252

golf courses should have no negative impacts on the surrounding environment. This information would be useful to those in decision and policy-making positions, as well as to the general golf course community, as a means to falsify erroneous claims regarding the negative effects of Roundup application.

References

1. Su, Y. *Ph.D. thesis, Kansas State University*, Manhattan, KS, 2002.
2. Heier, T.J. *M.S. thesis, Kansas State University*, Manhattan, KS 2004.
3. USEPA. RED (Reregistration Eligibility Decision) Document: *Glyphosate*; EPA-738-R-93-011; U.S. Environmental Protection Agency, Office of Pesticide Programs. 1993. www.epa.gov/oppsrrd1/REDs/factsheets/0178fact.pdf (accessed July 2004).
4. USEPA. Office of Ground Water and Drinking Water, Washington DC. *Technical Factsheet on: Glyphosate*, 2002, http://www.epa.gov/OGWDW/dwh/t-soc/glyphosa.html (accessed July 2004).
5. Acquavella, J.F.; Weber, J.A.; Cullen, M.R.; Cruz, O.A.; Martens, M.A.; Holden, L.R.; Riordan, S.; Thompson, M., *Hum. Exp. Toxicol.* **1999**, *18*, 479-486.
6. WHO (World Health Organization). *Glyphosate, Environmental Health Criteria*, 1994, No. 159.
7. Racke, K.D. In *ACS Symposium Series 743: Fate and Management of Turfgrass Chemicals.* J.M Clark, J.M. and Kenna M.P., eds.; 2000, p. 45-65.
8. Talbot, A.R.; Shiaw, M.; Huang, J.; Yang, S.; Goo, T.; Wand, S.; Chen, C.; Sanford, T.R. *Hum. Ex. Toxicol.* **1991**, *10*, 1-8.
9. EXTOXNET (Extension Toxicology Network). *Pesticide Information Profile: Glyphosate,* 1994, http://pmep.cce.cornell.edu/profiles/extoxnet/dienochlor-glyphosate-ext.html, (accessed July 2004).
10 U.S.D.A Forest Service. *Glyphosate, Pesticide Fact Sheet.* Prepared by Information Ventures, Inc, 1995, http://infoventures.com/e-hlth/pesticide/glyphos.html (accessed July 2004).
11. Stevens, J.T.; Sumner, D.D. *Herbicides* in Handbook of Pesticide Toxicology, Hayes and Law eds. Academic Press, NY. 1991, Vol. 3.
12. U.S. EPA. *Health Advisory*, Office of Drinking Water, 1987.
13. U.S. EPA. *Pesticide Tolerance for Glyphosate.* Federal Register, Vol. 49, No. 57, p. 8739-40. 1992.
14. NPTN (National Pesticide Telecommunications Network). *Glyphosate Technical Fact Sheet.* 2000. http://npic.orst.edu/factsheets/glyphotech.pdf.

253

15. Franz, J.E.; Mao, M.K.; Sikorski, J.A. *Glyphosate: A Unique Global Herbicide*, American Chemical Society, 1997, Vol. 4, p. 65-97.
16. Ghassemi, M.; Farge, L.; Painter, P.; Quinlivan, S., Scofield, R.; Takata, A. *Environmental Fates and Impacts of Major Forest Use Pesticides*, p. A-149-168, U.S. EPA, Office of Pesticides and Toxic Substances. Washington DC, 1981.
17. Laerke, P.E. Foliar. *Pest. Sci.*, **1995**, *44*, 107-116.
18. DellaCioppa, G., Bauer S.C., Klein B.K., Shah D.M., Fraley R.T., Kishore G.. *Proc. Natl. Acad. Sci. USA*. 1986, Vol. 83. 6873-6877
19. Ahrens, W.H. *Herbicide Handbook*, 7th Ed., Weed Science Society of America, Champaign, IL, 1994, p. 149-152.
20. Kenna, M.P.; Snow, J.T. In *ACS Symposium Series 743: Fate and Management of Turfgrass Chemicals*. J.M Clark, J.M. and Kenna M.P., eds.; 2000, p. 2-35.
21. Schuette, J. *Environmental Fate of Glyphosate. Environmental Monitoring & Pest Management*, California Department of Pesticide Regulation, 1998, www.cdpr.ca.gov/docs/empm/pubs/fatememo/glyphos.pdf. (accessed July 2004).
22. Kollman, W.; Segawa, R. *Interim Report of the Pesticide Chemistry Database, Environmental Hazards Assessment Program*, California Department of Pesticide Regulation, 1995.
23. Eriksson, K.E. Roundup®, *Weeds and Weed Control*, 1975, 16, J5-J6.
24. Rueppel, M.L.; Brightwell, B.B.; Schaefer, J.; Marvel, J.T. *J Agric. Food Chem.* **1997**, *25(3)*, 517-528.
25. Monsanto Company. *Backgrounder: Formaldehyde is not a Degradate of Glyphosate*. 2002. www.monsanto.com/monsanto/content/products/productivity/roundup/ gly_formald_bkg.pdf (accessed July 2004).
26. USEPA. *Federal Register.* 1997, Vol. 62, No. 70, p. 17723-17730.
27. Jonge, H.; Jonge, L.W.; Jacobsen, O.H. *Pest. Man. Sci.*, Abstract, **2000**.
27. Stratton, G.L.; Stewart, K.E. *Tox. and Water Quality*: A Inter. J., **1992**, 7, 223-236
28. Starrett, S. USGA Sponsored Research Annual Report, 2001.
29. Duble, R.L. *Zoysiagrass* Texas Cooperative Extension website, 2005, http://aggie-horticulture.tamu.edu/plantanswers/turf/publications/zoysia.html, (accessed February 2005).
29. Edwards, W.M.; Triplett, G.B.; and Kramer, R.M. *J Environ. Qual.* **1980**, *9*, 661-665.

Indexes

Author Index

Subject Index

A

Acetanilide, published measured thatch sorption coefficients, 191*t*

Acute toxicity, glyphosate, 239–240

Aerification, runoff deterrent, 143–144

Aminomethylphosphonic acid (AMPA)
  biodegradation of glyphosate, 242–243
  concentrations in samples from golf course, 246*t*
  See also Glyphosate

Ammonium nitrogen
  land use, losses by block, 29, 31*f*
  land use, mass loss in runoff, 29, 31*f*, 32–33, 40
  runoff depth and N mass losses, 27*t*

Application practices, inhibiting runoff, 146–147

Application rates, runoff deterrent, 141

Application timing, runoff deterrent, 140–142

Atlanta, GA
  fairway water input and runoff, 209*t*
  mean annual fairway runoff pesticides as percentage of annual application, 211*f*
  mean annual runoff of four pesticides, 210*f*
  one in ten year pesticide runoff events, 212*f*
  weather characteristics, 205*t*
  See also Pesticide runoff from turf

Atrazine, published measured thatch sorption coefficients, 192*t*

B

Bermudagrass, dense growth-form, and low nitrate and phosphate losses, 181

Bermudagrass buffers, attenuating runoff transport, 144–145

Biodegradation, glyphosate, 242–243

Block design, incomplete, statistical analysis, 97–98

Bromide tracers. *See* Clopyralid

Buffalo grass, runoff transport, 145

Buffer strips
  attenuating runoff transport, 144–145
  bioavailable phosphorus (P), 160, 162
  comparing total suspended solids (TSS) in prairie and turf, 158*f*
  establishment of, 153–154
  management, 154
  materials and methods, 153–155
  mean annual runoff volumes (2003, 2004), 156*t*
  mean P fractionation and loading (2003–2004), 161*t*
  phosphorus (P) loading from nonfrozen soil, 160
  proposal for reduction of runoff pollution, 152–153
  soluble P, 160
  surface runoff water quality, 155, 157
  surface water quality, 159–160, 162
  timing of erosion losses in turfgrass and native prairie vegetation, 159*f*
  TSS losses from prairie and turfgrass, 157–159
  water analysis, 154–155

266